Preventing the Greenlash

Also by Lorenzo Forni and published by Agenda

*The Magic Money Tree and
Other Economic Tales*

Preventing the Greenlash

How to Overcome Opposition to Green Policies

Lorenzo Forni

agenda
publishing

First published in 2024 by Agenda Publishing

Agenda Publishing Limited
PO Box 185
Newcastle upon Tyne
NE20 2DH
www.agendapub.com

ISBN 978-1-78821-781-1

British Library Cataloguing-in-Publication Data
A catalogue record for this book is
available from the British Library

Typeset by JS Typesetting Ltd, Porthcawl, Mid Glamorgan
Printed and bound in the UK by 4edge

Contents

Introduction

In 2023 and 2024, the world has recorded record-breaking temperatures, most of the time 1.5°C or more above preindustrial levels. Undeniably attributable to human activity and boosted by the natural phenomenon of the El Niño effect that warms the ocean surface, the world is warming at an unprecedented rate.

The emissions we produce through the burning of fossil fuels in the pursuit of economic activity – industrial production, consumption, travel, farming and heating our homes – are driving the planet towards extinction. Our modern lifestyles are to blame for accelerating the problem. Lifestyles that have been made possible through the economic system we have developed to enrich our society. The system of capitalism (the process of wealth creation through profit making), which underpins our economies, is organized through markets that balance demand and supply through price. Is it possible for this market capitalist system to now be used effectively to confront the climate crisis and mitigate the effects of global warming? Given Adam Smith's "invisible hand" theory, which posits that individual profit-seeking behaviour naturally steers economies towards the optimal distribution of resources, can we realistically expect this mechanism to drive a significant reduction in emissions?

Evidently, the market left to its own devices is insufficient to address the climate crisis. A fact that has been acknowledged by economists for some time. Indeed, major market failures are due to "externalities", which are the effects of market activity that impose a cost on a third party, such as ill health through breathing the polluted air emitted during the manufacture of a specific product. In scenarios where externalities exist, market outcomes in which the social cost inflicted is not included in the cost of the product keep prices low and demand high, leading to overproduction of both product and pollution. If, however, the social cost of these externalities is added to the production costs, it would drive up the consumer price reducing the amount demanded by the market. The manufacturer would then curb its level of production and consequently its pollution emissions, leading to a reduction in health-related problems in society. This internalization is not likely to be a voluntary act, per Adam Smith's dictum, but one that must be required by law.

With regards to greenhouse gas emissions that cause the externality of global warming, an effective way to integrate these costs into the production process, so that the producer "internalizes" their social cost, is through imposing a "carbon price" on emissions (either through a carbon tax or an emissions trading system, more on these later). The environmental impact then becomes a tangible part of the economic calculation for businesses and consumers.

Overproduction is at the heart of the environmental dilemma. While economic principles such as "externalities" help explain the mechanics behind our ecological footprint, the plain truth is self-evident: our capitalist societal norms of production and consumption are irreversibly damaging our natural surroundings. This issue transcends the emission of greenhouse gases; it includes the degradation of our water bodies, the depletion of green spaces and forests, the loss of biodiversity, and the encroachment upon natural habitats. For years, scientists have warned us about the unsustainable consumption of our "natural capital" – the abundance of resources the natural world provides in the forests, oceans, and entire ecosystems – that we deplete faster than it can regenerate, thus eroding its very existence.

The atmosphere is a vital component of our planet's natural capital, yet it is often mistreated as a communal wastebasket. Just as litter accumulates in an overflowing public rubbish bin – without directly affecting the personal property of those who contribute to the mess – greenhouse gases are relentlessly discharged into our shared atmospheric space. In a city, municipal workers may eventually empty the bins; however, there is no equivalent workforce to clean up the atmospheric pollution that results from our cumulative emissions. This predicament arises because of externalities: those who produce greenhouse gases typically shift the burden onto others and would incur costs if they chose to mitigate their emissions by adopting greener production methods. For example, a factory installing filters to clean its emissions does so at a substantial cost, with no immediate financial return. The benefits of reduced emissions – dispersed into the vastness of the atmosphere – are not directly perceivable by the individual or company bearing the expense. This disconnect between individual cost and collective benefit underscores the challenge of addressing climate change: the air one breathes and the local climate remain largely unaffected on a personal scale, despite the significant costs of implementing environmentally friendly practices.

However, externalities are not limited to industrial producers alone; it extends to every individual. Whether one is driving a car, heating a home, consuming goods, or using fertilizer, a common oversight persists: the failure to consider the environmental impact of the emissions related to the activities or products. These emissions have no immediate consequences for the individuals responsible. Yet, the cumulative effect of such externalities, generated by people worldwide through various activities, leads to a significant cost for society as a whole, namely global warming. It must be stressed that this excess production and emissions is not evenly distributed across the globe. Differences exists between regions with high per-capita consumption and emissions and impoverished areas, where the focus necessarily shifts from emission reduction to meeting the basic needs of nutrition, healthcare and education.

But overall, at the global level, our production and consumption levels exceed what our environment can sustainably manage, without due consideration for the resultant emissions we release into the atmosphere. By "exceed", I refer to an overuse that surpasses what our current technological solutions can handle without harming the environment and emitting large quantities of greenhouse gases. Ideally, we should aim to maintain – or even increase – our activities but through innovative methods that do not deplete our natural capitals and contribute to atmospheric pollution. Surprisingly, this objective is not as far-fetched as it might seem, given the technologies presently at our disposal.

Unfortunately, addressing the challenge of climate change is very difficult. Economic theory teaches us that when the actions of economic agents produce externalities, a series of sub-optimal behaviours are also generated. When individuals, like companies, have no incentive to reduce emissions because they derive no direct benefit, and indeed face costs to reduce them, a "free-rider" attitude prevails. If we leave it to others to act and reduce emissions, we gain the benefits from their reduction, but without having had to change our own habits or having to bear the costs of reducing our own emissions. The free-rider attitude makes impossible any cooperative behaviour in which we all make efforts to reduce emissions, which would be the best solution.

It should be added that, not only does reducing emissions not directly benefit those who do it, but also on an aggregate, even at the global level, reducing emissions tends to benefit mainly future generations and the few youngest among us who will survive until the end of this century. The concentration levels of greenhouse gases in

the atmosphere that have been reached so far will persist for centuries. Therefore, containing emissions today
means that the situation will not get worse, benefiting
the future. The challenge lies in the fact that today's
younger generations often lack significant political
influence. Future generations, not yet born and unable
to vote, remain abstract to us, making it difficult to prioritize their well-being. In short, we tend to discount the
distant future a lot, thus imposing another externality
on future generations, because we impose the costs of
our emissions on them. This is not to say that there
are not also immediate benefits from the reduction of
emissions, particularly related to the containment of
pollution in large cities, but even on pollution issues
the problem of free riding exists.

What actionable steps can we take? A central theme
of this book is the pivotal role of public policy and governmental intervention in addressing the climate crisis.
It is imperative that governments implement strategies that incentivize or mandate individuals to minimize externalities and prevent any opportunity for free
riding. These policies must be designed to align personal
choices with the broader, societal aim of reducing emissions on a global scale.

During the Covid-19 pandemic, the externality was
unmistakable – if I contracted Covid and visited a super-
market, I would potentially infect other people, who
would then spread the virus further. Infecting strangers
would not have directly impacted me (although, one
would hope, it would at least have provoked some
moral guilt) and is a clear externality. The government

intervened to stop the spread of the virus, requiring infected individuals to self-isolate. Later, with the advent of vaccines and their widespread adoption, free riding became apparent as some people felt that if others were vaccinated, they need not be. Here, too, authorities had to step in, requiring vaccination in certain contexts to counteract free riding and protect public health.

Undoubtedly, the measures taken during the pandemic – lockdowns and vaccines – evoke difficult memories for many. However, the worldwide shutdown demonstrated that climate change cannot be addressed with lockdowns and with the cessation of economic activity. Even at the height of the pandemic, when global activities slowed considerably, the reduction in emissions was minimal and fell short of the yearly decreases needed to achieve net zero emissions in the next two or three decades. During that time, energy consumption remained high, and consumer goods continued to be produced by factories worldwide.

Alas, we have no single remedy like a vaccine to counteract the climate crisis. The requisite policies for governments to enact to address climate change are multifaceted and complex. Both producers and consumers need to be steered towards practices that eliminate greenhouse gas emissions, and since there is little incentive to do so on an individual level, this effort must be collective, encompassing our whole society. It is essential to levy taxes on emissions so that emitters internalize the social costs they inflict; to champion the adoption of renewable energy sources such as solar, wind, hydro and nuclear, as well as those still in development; and to invest in low-emission technologies for industries like cement, steel and fertilizer production, even when such advancements require substantial initial investment. Energy conservation must also be a priority, through enhancing the efficiency of our buildings and transport systems and promoting the principles of a circular economy.

Yet, none of these initiatives can succeed without a definitive mandate from us, the citizens and electorate, empowering our leaders to act swiftly on decarbonizing our economy. While seemingly straightforward, this imperative is laden with complexity. It is not solely about solving the technical challenges to make green technologies widely accessible, where considerable progress has been made. It is equally about forging a widespread societal agreement on managing the transition, including identifying the necessary and most effective measures for achieving decarbonization. On this front, we have much ground to cover. Consider the divergent perspectives of the United States, Europe and China, each adopting distinct approaches to this global imperative.

Even with the necessary technology, strategy and supportive policies for a green transition, the journey is far from straightforward. When fuel taxes spark public protests, household energy cutbacks meet resistance, or the automotive industry lobbies against emissions regulations, the path forward becomes murky. We are faced with a conundrum: how do we navigate the transition to sustainability when not all citizens are willing to shoulder the immediate costs for the long-term benefit of humanity? The tendency towards free riding – expecting others to act while we do not – is a formidable obstacle. So, what is the most effective approach? Is it feasible to impose these costs through coercive means, reminiscent of pandemic lockdowns, without compromising democratic principles? Furthermore, how can we expect emerging and developing countries, with their many pressing challenges, to prioritize the climate transition? It's evident that successful decarbonization requires not just policy and technology but also a shared societal commitment to enduring short-term sacrifices for the greater, long-term good.

How do we bridge the gap between the collective benefit of mitigating global warming – with its many associated costs – and the immediate advantages for individuals who reduce their emissions? Is it feasible to frontload the benefits from a sustainable future to offset the costs we incur today? Indeed, governments have the capacity to assume debt to finance the transition, with the expectation of repaying it as we move towards a greener economy. Can we thus bequeath to future generations a healthier planet alongside a manageable level of debt? It is a compelling proposition that many of our descendants would likely endorse, recognizing the exchange of some financial burden for a significantly more liveable world?

This book begins by acknowledging the scientific consensus: to prevent a progressive increase in global temperatures we must reduce greenhouse gas emissions to zero – a point that is examined in Chapter 1. Chapter 2 discusses the current economic frameworks for systematizing the climate crisis and for crafting policy solutions, highlighting their significant shortcomings. Chapter 3 scrutinizes the efforts of major world governments, evaluating their strategies and policies against the ambitious goal of curtailing emissions and illustrating the inadequacies therein. The complexities of forging a unified global response and the constraints of international climate agreements are laid bare in Chapter 4. Finally, Chapter 5 delves into the necessary policies for climate action, exploring strategies to make these measures both palatable and acceptable to the electorate.

Governments are uniquely positioned to address externalities and prevent free riding; they alone have the authority to mobilize the additional resources required to mitigate the costs associated with the transition to a greener economy. By harmonizing individual motivations with the collective welfare and transforming the nebulous concept of "long-term benefits" into something immediate and personal, we can enhance public support for bold climate policies.

1

Where we are in the climate transition

"Adults keep saying we owe it to the young people, to give them hope, but I don't want your hope. I don't want you to be hopeful. I want you to panic. I want you to feel the fear I feel every day. I want you to act. I want you to act as you would in a crisis. I want you to act as if the house is on fire, because it is".

Greta Thunberg, speech at World Economic Forum meeting, Davos, 2019.

Keep calm and carry on

On 7 June 2023, all flights out of LaGuardia airport in New York were cancelled due to the thick smoke that enveloped the city. Wildfires in the Canadian province of Quebec triggered air quality warnings down the Eastern Seaboard of North America. Over the course of that summer a series of fires devastated 5 per cent of Canada's forests. This exemplifies the daily influx of news detailing environmental catastrophes linked to global warming, heightening anxiety and uncertainty about what is actually happening. Undoubtedly, many of us are grappling with the same questions: What is the current increase in global temperatures? How much will temperatures rise in the coming years? Are human actions entirely responsible for these changes? And what measures can we implement to mitigate the most severe consequences?

While it is undeniable that recent natural catastrophes – from wildfires and floods to hurricanes, heatwaves and droughts – have increased public awareness of climate issues, levels of public concern vary. For instance, in 2023, approximately 80 per cent of European Union (EU) citizens viewed climate change as a serious threat, whereas in the United States (US) the proportion was 40 per cent.[1] Despite rising awareness of the potential threats posed by global warming, robust public demand for mitigation remains absent. Rather than national and international collaborative actions to implement emissions-reducing strategies, we see divergence and debate over who should bear the burden of such initiatives. This stalemate is understandable given the multifaceted challenges, especially concerning the equitable distribution of the green transition costs, that impede swift action. Before a deeper investigation of these intricate matters in the succeeding chapters, it is first imperative to assess the current state of greenhouse gas (GHG) emissions and the potential repercussions of their future trajectories.

A brief note for clarity: the aim here is to address the problems and to propose solutions grounded in evidence and reasoning. Among the many different individual perspectives, my approach aims to be objective. I have no ties to companies with environmental footprints or with organizations focused on climate issues and I do not occupy a government position that might influence my views. I am an economist whose research extends to examination of these urgent issues. My motivation is simply to explore the complexities of global warming and to discern effective policy measures to mitigate its consequences.

Some key principles

The "greenhouse effect" is a natural process, which causes the warmth from the sun to be trapped in the lower atmosphere and ensures that the temperature across most of our planet remains at levels that allow human beings to survive. It works as follows: the sun emits energy in the form of visible light and other types of electromagnetic radiation. This energy travels to the earth where some of this incoming solar radiation is absorbed and works to warm the planet. As the earth's surface warms, infrared radiation (or heat) is emitted back into the atmosphere. This energy type is invisible but can be felt as heat. The earth's atmosphere includes gases, collectively described as GHGs, which have the ability to trap some of this outgoing infrared radiation. The most common GHGs are carbon dioxide (CO_2), methane (CH_4), water vapour (H_2O), nitrous oxide (N_2O) and fluorinated gases. These gases allow the passage of the incoming solar radiation but prevent some of the outgoing infrared radiation from leaving the atmosphere.

This trapped infrared radiation causes warming of the earth's atmosphere. Under stable conditions, the amount of energy the earth receives from the sun is roughly equal to the amount of energy it emits back into space, which means that, over the long term, the earth does not continue getting hotter or colder. The presence of GHGs keeps the earth surface temperature much warmer than it would be without them.

Human activities, such as burning of fossil fuels, deforestation and industrial processes, are increasing the concentration of GHGs (especially CO_2) in the atmosphere. This enhanced greenhouse effect is causing more heat to be trapped, which is increasing global temperatures and producing what we call global warming.

Scientists agree about three aspects of the enhanced greenhouse effect which are fundamental to our understanding of the implications of global warming:

1. The temperature rises are due to increased GHG *levels* in the atmosphere. The difference between the "level" and "flow" of GHGs is crucial. Level refers to the total concentration of GHGs in the atmosphere at any given time; flow refers to the volume of GHGs added or removed in a specific period (e.g., annually). The problem concerns not just the volume of the additional emissions being pumped

into the atmosphere annually, but their cumulative level and their increase over time. GHGs, especially CO_2, can persist for centuries or even millennia.

2. The increases in emissions and, therefore, in atmospheric concentrations is caused mainly by humans. Human activity has been the primary driver of this alarming increase. Industrialization, deforestation and the burning of fossil fuels have significantly ramped up CO_2 and other GHG amounts in the atmosphere, disrupting the natural balance and causing an enhanced heat-trapping effect.

3. GHG concentrations in the atmosphere have reached a level where their further increase will lead to a significant rise in average global temperatures. Historical data and ice core samples indicate that current GHG and particularly CO_2 levels are higher than for hundreds of thousands of years. Continued emissions mean concentrations will increase – exacerbating the greenhouse effect. In the absence of significant interventions, these average global temperatures will continue to rise with huge negative implications for our planet.

Although there are certain nuances and specifics that might be debated, the scientific community acknowledges these foundational points, and the 2021 Sixth Report of the United Nations (UN) Intergovernmental Panel on Climate Change (IPCC) defines them as "unequivocal" (IPCC 2021). Understanding these basics is the first step towards appreciating the magnitude of the challenge at hand and the need for urgent and coordinated global

action. It has taken years of rigorous scientific explora-
tion to arrive at these conclusions, which should pave
the way to agreement among and action by governments. The landmark 2015 Paris Agreement states that
the relentless rise in GHGs cannot persist indefinitely
without an uncontrolled escalation in atmospheric
GHG concentrations and unsustainable global temperature hikes. Since there is a thermal threshold beyond
which human survival becomes precarious, a cap on
future emissions is essential. In essence, emissions must
be reduced to a *net zero balance*. This goal has profound implications for global economic and societal
trajectories, but achieving net zero is imperative.

The immediate challenge is operationalizing the net
zero objective. To prevent temperature rises of more than
2°C, beyond which adverse climatic events are likely
to escalate uncontrollably, renowned climatologists,
especially the authoritative IPCC – the UN's primary
climate change scientific body – propose its achievement by 2050. The more recent 2022 IPCC assessment
recommends that, for optimal safety, temperature rises
should be capped at 1.5°C relative to pre-industrial
levels (IPCC 2022). There is a large degree of consensus in the scientific community that exceeding 1.5°C or
2°C degrees might trigger unpredictable climate events
or reach tipping points,[2] although the uncertainty on
exactly what could happen is very high and is the object
of continuing scientific research.

The bathtub analogy

The intricacies of the climate change problem can be simplified using the household example of the bathtub. Imagine walking into the bathroom to find the bath brimming with water and the tap still running. Your immediate concern is to prevent a flood by turning off the tap or pulling out the plug. The first stops the water level from rising but has no effect on the water level in the bath. To reduce the water level the plug must be removed.

In this analogy, the water in the bath is the existing level of the GHGs in the atmosphere; the running tap is the additional annual GHG emissions. The additional gases raise the overall level (or concentration) of GHGs. As GHG concentrations increase, the greenhouse effect intensifies, causing global temperatures to rise. If GHG concentrations pass a certain threshold, there is a risk that temperatures will rise by more than 2°C, which scientists forecast will have unpredictable and inevitably detrimental climate outcomes. In our bathtub scenario, this is akin to the damage caused by the bath overflowing.

To extend this analogy, we need to reduce the water level in the bath or decrease the concentration of GHGs in the atmosphere. In the former case, this requires only removal of the plug; in the latter case, there is no plug and these excess GHGs could linger – sometimes for centuries, just as if the bath had no plug or drain hole. In the bathtub example, theoretically, we could bail out the water, although this would be a tedious and slow process. In the case of the atmosphere, the solution is extraction and storage of atmospheric CO_2, which would require efficient and scalable technologies that are not currently available. This underlines the importance of curtailing emissions to near-zero levels to avoid an "overflowing bath" and its inevitable dire consequences.

The net zero objective

When discussing net zero emissions, we need to consider certain nuances. It is important to note that we use the term "net zero" and not just "zero". Many readers will recognize that the distinction stems from the fact that achieving "absolute zero" emissions by 2050 would be probably impossible, given the current technology landscape. In the bathtub analogy, we might never be able to turn off the water completely. This means that by 2050, it is inevitable that certain activities will continue to produce GHGs. Therefore, we need to find a balance, based either on trapping and storing these emissions underground (a technique known as carbon capture and storage) or developing futuristic innovations to allow the capture of atmospheric CO_2, such as carbon sequestration or carbon removal to enable negative emissions. Currently, we have no efficient means of extracting CO_2 from the atmosphere; existing methods require energy which, paradoxically, generates more emissions than these methods are able to extract.

The broader discourse on climate warming proposes the objectives of net zero by 2050 and capping global temperature rises at 1.5°C. Some critics argue that fixating on precise benchmarks could be counterproductive. Indeed, the IPCC does not propose a definite threshold, but suggests scenarios based on probabilities. Drawing on comprehensive research conducted by international expert teams, the IPCC suggests that achieving net zero CO_2 emissions by 2050 would make preventing temperatures from escalating beyond 1.5°C a "high probability" event. However, meeting the 2050 target would not guarantee that temperature rises would be confined to 1.5°C. On the other hand, there is a very small chance of not going beyond 1.5°C even were net zero not achieved until after 2050.

Those opposed to the imposition of strict policies to reduce emissions, point to the lack of certainty that temperatures will remain contained even were we to reach net zero by 2050 and, also, that temperatures may not continue to rise excessively. However, the main argument rests on the very high costs of reducing emissions that must be borne today, while the gains, in terms of containing temperatures and reducing adverse natural phenomena, could be limited and would emerge only in the distant future. In short, why should we incur high economic costs today to achieve a small reduction in emissions in order to reap modest gains tomorrow?

However, these arguments neglect a fundamental truth that, eventually, net emissions must reach zero. This is a necessary not a debatable outcome. Those arguing against swift emissions-reduction measures tend, typically, to focus on a timeline up to 2100 and specific models that predict contained economic and social repercussions from rising temperatures at the end of the century. Although there is consensus about the inevitability of needing to achieve net zero emissions, discussion about the optimal pace of the changes needed to achieve this goal is less than agreed.

Delaying action could have irreversible environmental and economic consequences, while hasty measures would add to the strain on current economies and, potentially, could affect existing social structures and future growth. At the crux of the issue is finding a balance between immediate economic concerns and long-term environmental imperatives. This is the main focus of this book. However, first, we need to consider the current state of global emissions and countries' stated goals and plans to reduce them.

Emissions are growing

Despite announcements by numerous governments of their ambitious emissions reduction goals, GHG emissions are increasing continuously.[3] Despite a temporary decline in 2020, due to the global Covid pandemic (see Figure 1.1), emissions have increased steadily since the Second World War. Current CO_2 emissions range between 35–40 billion tonnes each year, and account for roughly 75 per cent of world GHG emissions, followed by methane (17 per cent), nitrous oxide (6 per cent) and a small volume of fluorinated gases.

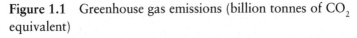

Figure 1.1   Greenhouse gas emissions (billion tonnes of CO_2 equivalent)

Note: Greenhouse gas emissions include carbon dioxide, methane and nitrous oxide from all sources, including land-use change.

Source: Our World in Data.

Table 1.1 depicts emissions based on their sources. A large proportion is linked to energy consumption, involving sectors such as industry, transportation and housing.[4] Approximately 6 per cent are due to fugitive emissions or accidental leakages from industrial infrastructures and energy production facilities. Agriculture, livestock farming, and forestry are notable contributors and several industrial processes, especially cement and chemicals production, emit GHGs. Use of fertilizers results in the release of nitrous oxide (N_2O), a gas whose greenhouse effect is more potent than that of CO_2. Thus, it is clear that most human activities produce GHG emissions (see Gates 2021).

Figure 1.2 provides a detailed emissions breakdown, for the ten nations with the largest carbon footprints. In 2023, China was in pole position, with a significant surge in emissions in line with its remarkable economic

Table 1.1 Global greenhouse gas emissions by sector (2016)

Sector	Tier 1 (%)	Tier 2 (%)
Energy	**73.2**	
Transport		16.2
Energy in buildings		17.5
Energy in industry		24.2
Energy in agriculture and fishing		1.7
Unallocated fuel combustion		7.8
Fugitive emissions from energy		5.8
Industry	**5.2**	
Cement		3.0
Chemical		2.2
Agriculture, forestry and land-use	**18.4**	
Livestock and manure		5.8
Rice cultivation		1.3
Agricultural soils		4.1
Crop burning		3.5
Deforestation		2.2
Cropland		1.4
Grassland		0.1
Waste	**3.2**	
Landfills		1.9
Wastewater		1.3

Source: Our World in Data.

growth since the early 1980s. Ranked second was the US, which, interestingly, has seen a decrease in overall emissions since the early 2000s as a result of technological advancements and a transition from coal to natural gas. India's fast-growing population and position as a major energy importer ranks it third. Other major contributors include prominent oil producers, such as Indonesia, Iran, Russia and Saudi Arabia. If we consider cumulative emissions since the beginning of the Industrial Revolution, the narrative shifts. Historically, the industrialized nations, primarily the US and European countries, have been the main culprits in terms of accumulation of GHGs in our atmosphere, but China is fast catching up to the levels cumulated by the advanced economies (see Figure 1.3).

Figure 1.2 Annual CO_2 emissions from fossil fuels and industry (excluding land-use change)

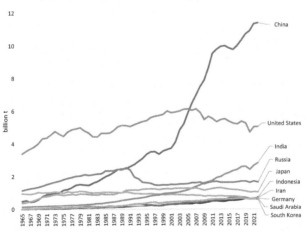

Source: Our World in Data.

Figure 1.3 Cumulative CO_2 emissions produced from fossil fuels and industry (excluding land-use change) since the first year of recording, measured in tonnes

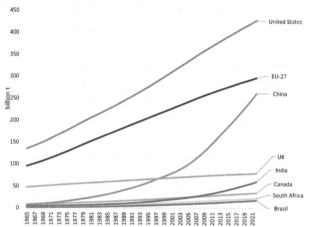

Source: Our World in Data.

The Kaya Identity

The Kaya Identity is a framework for understanding the factors that contribute to the growth of CO_2 emissions. It breaks down carbon dioxide emissions into four factors: population size, GDP per capita, energy intensity of GDP, and carbon intensity of energy.

The equation is written as

$$CO_2 \text{ Emissions} = \text{Population} \times \frac{\text{GDP}}{\text{Population}} \times \frac{\text{Energy}}{\text{GDP}} \times \frac{CO_2}{\text{Energy}}$$

When you multiply these four factors together, the units cancel in such a way that you are left with the total CO_2

emissions. The Kaya Identity is thus a way to factorize CO_2 emissions, the most significant GHG for climate change that is produced by human activity, into demographic, economic and technological components.

To explain each component of the equation:

Population is the total number of people inhabiting the planet. As population increases, more energy is generally consumed, leading to higher CO_2 emissions.

GDP per capita (GDP/population) is a measure of the average individual economic output and is measured in dollars per person. When it increases, it typically means that individuals are wealthier and consume more, which often results in increased energy use and hence more CO_2 emissions.

Energy intensity (Energy/GDP) assesses how much energy is used to produce a unit of GDP. It is usually measured in joules per dollar (or another energy per economic output unit). A decrease in energy intensity means that less energy is needed for the same economic output, potentially leading to lower emissions if the energy sources are constant.

Carbon intensity (CO_2/energy) measures how much CO_2 is emitted per unit of energy consumed. Lower carbon intensity means that the energy being used is cleaner, i.e., it produces less CO_2 for the same amount of energy, which is typically achieved by shifting to renewable energy sources.

Figure 1.4 shows that CO_2 emissions have increased significantly, by over 225 per cent, since 1965. This is a result of a combination of factors but notably driven by an increase in GDP per capita and by a rise in population. GDP per capita has been increasing steadily, reflecting economic growth worldwide. Economic development usually corresponds with increased energy demand and consequently emissions, although the rate of emissions growth can be mitigated by improvements in energy and carbon intensity. Population increase contributes to the rise in CO_2 emissions. More people means a greater demand for energy, goods and services, often resulting in more emissions. Energy intensity has decreased by over 40 per cent, which is a positive trend. This indicates that economies are becoming more efficient in how they use energy to produce economic output. Carbon intensity has also been decreasing, by about 15 per cent, although not as sharply as energy intensity. This indicates a gradual shift towards less carbon-intensive energy sources, but not at a pace that is able to offset the increase in emissions from GDP and population growth.

The combined effect of these factors demonstrates that although there has been some progress in energy efficiency and cleaner energy sources, the overall trend is still an increasing trajectory of CO_2 emissions, driven largely by economic and population growth. This highlights the challenge of decoupling economic development from carbon emissions in order to achieve climate goals.

The Kaya Identity offers a valuable framework for deconstructing the primary factors driving carbon emissions, supporting a strategic approach to mitigating them. Throughout this book, we shall elucidate the

Figure 1.4 Percentage change in the four parameters of the Kaya Identity, which determine total CO_2 emissions

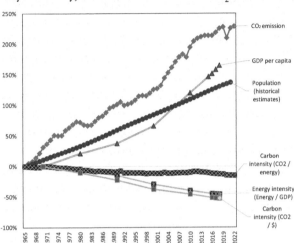

Note: Emissions from fossil fuels and industry are included; land-use change emissions are not included.

Source: Our World in Data.

most impactful actions that can be taken in reducing both the energy and carbon intensity of our economies. To diminish carbon intensity, an imperative shift is necessary – away from fossil fuels and towards low-carbon energy sources – to bolster the role of renewables. In relation to energy intensity, a suite of strategies designed to minimize energy consumption for each unit of economic output will be explored. These include efforts to curtail energy wastage and the implementation of innovative practices encompassed by the "circular economy". These topics will be explored comprehensively in Chapter 5.

Addressing emissions through the lens of demographics presents intricate challenges, especially considering United Nations projections of plateau in the global population at approximately 10 billion by the year 2050. Even more controversial is the debate over whether to limit or reduce GDP per capita growth as a means of containing emissions. Chapter 2 will delve into why suppressing economic growth is neither a viable nor effective strategy for emission reduction.

As we navigate the complex discussions surrounding policy measures for emission reduction, the Kaya Identity remains an accessible and potent tool, serving as a constant reference point amidst the multifaceted aspects of climate policy debates.

The three gaps

In autumn 2023, the UN Environment Programme (UNEP) published its *Emissions Gap Report 2023*, which offers a stark reminder of the challenges ahead. It states that, in 2022, global overall GHG emissions soared to 57.4 billion tonnes of CO_2 equivalent,[5] marking a return to the pre-pandemic upward trajectory. The UNEP report highlights that the majority – approximately two-thirds – of these emissions, stem from the use of fossil fuels and industrial activities. Alarmingly, the gap between the current trajectory and the target emissions for 2030 remains largely unbridged. While countries' announced Nationally-Determined Contributions (NDCs)[6] would, if fully implemented, reduce emissions by 7 per cent by 2030 (compared to 2019 levels) (Fransen *et al.* 2023), this falls significantly short of the 42 per cent reduction required to maintain a chance of limiting global warming to 1.5°C. The UNEP assessment is unequivocal: if we persist with the status quo, we will be on track for an average global temperature rise of 3°C by the end of this century. The report underscores a sobering reality, noting that even in the most optimistic scenarios, the likelihood of constraining global warming to 1.5°C is a mere 14 per cent. It is important to point out that, as along as net emissions remain positive, global temperatures will keep rising even past the end of this century. Therefore, warming estimates given for 2100 are not an upper bound on possible temperature increases, but simply the expected values at a particular point in time – the end of the century – of a rising temperature trend.

The growth in emissions belies the many national pledges to reduce them. For example, it would be nice to believe that the commitments in the 2015 Paris Agreement would solve our environmental woes; however, despite considerable progress they are nowhere near sufficient. Although many countries have committed to curbing their emissions within the scope of the Paris Agreement and their detailed NDCs, the reality is complicated. Essentially, the NDCs are formal statements, outlining the individual country's pledges, related to the actions to comply with the Paris Agreement. However, since the signing of the Paris Agreement, not all these NDCs have been translated into action. For instance, at COP26 in Glasgow in 2021, it was clear that many of the Paris goals had not been achieved. When COP27 was held in Egypt in 2022, only 24 countries (none of them primary emissions producers) had updated their NDCs. These facts raise questions about how many of the noble intentions announced in the Paris Agreement have translated or will translate into actions.

On paper, at least, there is evidence of progress. The EU is increasing efforts to achieve a greener future. The RePowerEU plan, published on 18 May 2022, doubled down on the European Commission (EC) promise to curtail emissions by 2030 by 55 per cent from a 1990 baseline (European Commission 2022). This would achieve net zero emissions by 2050, a commitment anchored in the European Climate Law enacted on 29 July 2021. This commitment applies not just to a few European countries but is binding for all 27 EU member states. The EC RePowerEU plan proposes several strategies to foster energy efficiency, diversify energy production, hasten the integration of renewables, minimize reliance on fossil fuels and boost public sector investment. One of the EU's standout initiatives is the EU Climate Neutral and Smart Cities Mission, which seeks to transform 100 European cities into net-zero urban centres by 2030. It is intended that this pilot action, which will allow testing of innovative solutions to climate change, will be extended to include all European cities by the middle of the twenty-first century.

Other countries are also implementing assertive actions against climate change. In spring 2022, the UK (independent of the EU, post-Brexit), committed to an impressive 78 per cent reduction in emissions by

2035 compared to the baseline 1990 levels. In the US, under President Biden's leadership, an ambitious albeit non-binding target was announced, to reduce net GHG emissions by 50–52 per cent by 2030 against the benchmark of 2005 levels. This commitment was reinforced by the enactment of the Inflation Reduction Act, hailed by many as the most significant step towards addressing climate change in America's history. The legislation and its myriad strategies to boost domestic manufacturing of clean energy technologies (e.g., solar, wind), carbon capture solutions and green hydrogen production, certainly packs a punch. The journey to achieving a more sustainable global environment is not limited to these major players; several nations have made remarkable progress, based on phasing out coal and introducing natural gas as their power generation source. Although natural gas produces fewer emissions than oil or coal, it is a fossil fuel and therefore is not a long-term solution.

However, although current global efforts to address climate change may be commendable, they are not enough to claim that global emissions are on a downward trend. Also, going forward, the journey to the realization of the ambitious targets set in the Paris Agreement, is complicated. Although significant, present commitments fall short of meeting the IPCC's target of limiting global warming to 1.5°C by the end of this century. Current projections show that, even were all these commitments fulfilled, world temperatures could rise by more than 2°C by 2100 – well beyond the 1.5°C IPPC target (see, e.g., Climate Action Tracker 2023). Achievement of the 1.5°C goal, according to the IPPC, will require global emissions to reduce by at least 43 per cent by 2030 relative to 2019 levels. This is in stark contrast to the reduction pledged in the existing NDCs.

Before COP27 in November 2022, a UN Framework Convention on Climate Change (UNFCCC) report assessed that current climate strategies could lead to global warming in the range 2.1–2.9°C by year 2100. If the commitments in the NDCs updated post-2021 are delivered, they will reduce global warming only slightly compared to what was agreed at COP26 in 2021, leading to an average warming of 2.5°C by 2100. Furthermore, UNEP states clearly that current policies do not offer a "credible pathway to 1.5 degrees". The World Resources Institute suggests that nations will have to amplify their NDC emissions reduction ambitions sixfold to align with the 1.5°C target. A significant portion (over 85 per cent) of such reductions should come from the doubling of efforts in existing sectors and technologies. This highlights the urgent need to refine and optimize current strategies, rather than pinning hope on new sectors and breakthrough technologies.

Given all this evidence, the global climate change challenge can be summarized, in terms of what I call the "three gaps problem", in which each gap refers to a distinct shortcoming.

1. *Implementation gap*: This is the difference between what countries planned to do and what they are actually achieving. Execution of ambitious climate strategies and roadmaps often lags, hindered by

inadequate policy mechanisms, lack of resources, political shifts, and other factors.

2. *Commitment gap*: There is a clear disparity between countries' current plans and their international commitments. Many nations announced admirable climate targets, but their strategies and stated milestones are not in line with these targets and will fail to achieve the desired outcomes.

3. *Ambition gap*: Global commitments in response to the dire need to limit global warming to 1.5°C are currently inadequate. Even if all nations were to fulfil their stated commitments, this collective effort would not prevent significant warming.

Addressing these three gaps will require concerted efforts from all governments, businesses, communities, and individual people. Bridging these gaps, and tackling the looming climate crisis seriously, demands a holistic and united approach.

Should uncertainty deter action?

Although there is clear and significant uncertainty about the precise effects of climate change, this uncertainty should not be used as an excuse not to respond. Were we to achieve net zero before 2050, this could exceed our climate goals and, potentially, might result in temperature increases below the 1.5°C threshold – a favourable although unlikely outcome. Conversely, if we should fail to achieve net zero by 2050, we risk significant temperature hikes that will likely foster a cascade of adverse environmental events. It must be understood that once temperatures rise, the situation is essentially irreversible since GHGs can remain in the atmosphere for hundreds and even thousands of years. If the net zero objective is not achieved, in the worst-case scenario the continual rise in temperatures could threaten human existence.

At the crux of this issue is the "precautionary principle", which suggests that, in the face of – especially irreversible – potential harm, it is wise to err on the side of caution. The risk of perhaps doing more than is strictly necessary is a much better option than the alternative of risking the very future of humanity. So, should uncertainty prevent action? Given the stakes, the answer would seem to be no.

Uncertainty about whether achievement of net zero by 2050 would stave off the direst consequences of climate change continues to loom large. Given current alarming trends, such as rising temperatures and an uptick in extreme weather events (e.g., droughts, wildfires, floods, tornadoes), the feeling that the 2050 net zero target might not be sufficiently aggressive is becoming more widespread.

It might be that we will need to accelerate our efforts. That is, if we manage to hit net zero only by 2050 or even later, we will be flirting with very substantial risks. Also, even if we assume a very unlikely low probability of disastrous outcomes, the potential ramifications of delaying net zero efforts are likely to be enormous: the expected costs of an extreme event, based on multiplying event probability by its subsequent impact, can be exorbitant. From a risk management perspective, minimizing exposure to extreme weather events that exact hefty economic and human tolls, would be the wisest course and emphasizes the urgency for expeditious progress towards the net zero target. In reality, the responses from governments around the world appear to diverge from the precautionary principle and a risk management perspective. Although several countries have articulated their commitment to curbing emissions, implementation of solid policies to achieve this tends to be absent.

The events of 2022–24, such as the steep rise in energy prices and the geopolitical ramifications of Russia's invasion of Ukraine and the war in the Middle East, took over the headlines and eclipsed the urgency of the green transition. In their efforts to reduce reliance on Russian energy, European governments have resorted to more affordable alternatives, including partial reversion to fossil fuels including oil and coal. This is a further setback to the imperative to slash emissions. Similarly, the surge in energy prices during 2021 and 2022 was so pronounced that several European administrations were forced to provide significant fiscal subsidies to stabilize prices. This action is in stark contrast to the demands of climate conservation, which call for higher CO_2 prices that, inevitably, lead to increased energy costs.

It is clear to most that achieving net zero is vital for societal well-being. The precautionary principle underlines the wisdom of erring on the side of caution and that it is more prudent to do more rather than less – especially considering the irreversibility of potential temperature escalations. These guiding principles lead to a crucial takeaway and the core theme and aim of this book: the pressing need to expedite the journey towards net zero which requires accelerated climate change efforts because the current trajectory falls well short of this net zero objective.

2

The green transition and growth

"The problems of the world cannot possibly be solved by sceptics or cynics whose horizons are limited by the obvious realities. We need men who can dream of things that never were and ask why not."

John F. Kennedy, address before the Irish Parliament, 28 June 1963.

The alarming rise in emissions and the challenges faced by countries in trying to curb them begs the question: What is holding governments back? Are the political and economic impediments to a zero-emissions future so great that any attempts to reach it are futile? In this chapter, I tackle a key question related to the transition to a greener economy, namely, does it make economic sense? Will there be gains for economic growth and job opportunities or are we merely staring down the barrel of huge financial costs and, potentially, compromising current living standards for a cleaner environment?

To begin with, I want to dispel a commonly held belief that reducing our carbon footprint inevitably requires a decline in our economic activities – an idea that has been described as "degrowth theory". In order to challenge the idea of degrowth, I explain the medium- to long-term concrete and substantial economic benefits we stand to gain from transitioning to an emissions-free future. However, it is necessary to acknowledge that the path towards a sustainable future is fraught with significant challenges. In this chapter, I explore the most salient obstacles, and I shall discuss strategies to overcome them in the following chapters.

Is degrowth the solution?

A pressing concern for many is the sustainability of the existing capitalist system, which describes an economy characterized by the private ownership of the means of production and their operation for profit. Individuals and companies own businesses and property, and the main goal of companies is to generate profits. Market forces, primarily supply and demand, determine the prices of goods and services, fostering a competitive environment in which businesses compete for customers. This competition is believed to drive innovation and efficiency. The role of governments in this system typically is limited and focuses mainly on enforcing contracts and property rights, with the idea that it is the market that is the primary regulator of economic activity.

The capitalist system's foundational emphasis on profit generation and reinvestment means that it is inherently reliant on continuous growth. Businesses and economies must grow to remain viable and competitive. Growth allows for increased profits, which are required to attract investment, fund innovation activity, and expand operations. Expansion of operations is not just a measure of success but is a requirement for survival in competitive markets where stagnation can lead to obsolescence. Also, under capitalism, economic growth is often seen as the means to improve living standards and create job opportunities. However, this incessant need for growth can lead to problems, such as resource depletion and environmental damage, which highlights the complex interplay between the economic system and sustainable development.

Let me be clear on this point: even zero growth would harm the environment and be unsustainable. Zero growth means that every year we would produce the same amount of goods and services. This, at least in advanced economies, which tend to have stable or decreasing populations and high living standards, might even be seen as desirable. But if we keep the current levels of production with the current technologies, we would still be emitting unsustainable amounts of GHG every year and continue depleting our natural capitals. Even zero growth will not save us from global warming.

The natural question to ask ourselves is whether we should uphold the capitalist system as the foundation of our society, or do we now require a fundamental shift away from it? Surely, if the current economic system is depleting the earth's natural resources and increasing global warming, its long-term viability is questionable. This predicament is tied to the broader perspective that perpetual growth is untenable. In its current form, continuous expansion is eroding the ecosystem that supports us and is leading, potentially, to the downfall of humanity.

A prevailing concern, voiced by many, is the harm wrought by our current "capitalist" trajectory, in depleting and harming the planet's natural capital and biodiversity. "Natural capital" refers to the world's stock of natural resources, which includes geology, its soil, air, water and its living organisms. Many of these resources are exploited for production: forests harvested for timber or converted for agriculture; oceans used for fishing and international shipping, which cause pollution; and water quality threatened by industrial effluents and untreated sewage, while terrestrial regions suffer contamination from the discharge of industrial waste. Also, the atmosphere is a significant part of natural capital. Production practices that produce GHGs are disrupting the natural balance of the atmosphere and increasing global warming.

There is a clear and similar urgency related to limiting GHGs to preserve the atmosphere and regulating human activities to safeguard forests, oceans, freshwater sources and more. This urgency is demonstrated by numerous studies that concur that our productive endeavours are jeopardizing these natural assets (Dasgupta 2021). In addition, biodiversity, which is tied intrinsically to the health of natural capital, faces significant threats. As we damage and consume the earth's natural reserves, we imperil countless living species, which underlines the pressing need to mitigate or even reverse depletion of the earth's natural capital. This will require striking a harmonious balance between protecting our natural capital and ensuring economic growth. In essence, the solution is sustainability.

Sustainable economic progress refers to maintaining economic growth while ensuring responsible use of natural resources to preserve them for future generations. It involves the creation of economic value in a way that does not deplete the world's natural capital. The concept of sustainable economic progress emphasizes efficient use of resources, waste reduction, and innovation to provide both economic and environmental benefits. The goal is to maintain an equilibrium, such that economic activities do not harm the environment or exhaust resources faster than they can regenerate and ensuring long-term economic health and an ecological balance.

Merely limiting consumption and production is neither a possible nor viable solution. Reducing consumption levels equates, essentially, with degrowth – an approach that seems both impractical and potentially ineffective, as evidenced by our experience of the Covid-19 pandemic. The 2020 peak in the pandemic and the resulting lockdowns and shutting down of society resulted in GDP downturns, ranging 20–30 per cent in some economies, but led to only modest (around 5 per cent) global emissions reductions. Thus, imposing drastic constraints on consumption and production is unlikely to produce the desired environmental outcomes.[7]

Given the need for average global emissions reductions to exceed 5 per cent every year over the next three decades to achieve net zero, cutting back is clearly not going to be enough. We must also consider the diversity in consumption and well-being among different nations. Should the same constraints be imposed on all countries despite their varying economic positions, or should there be convergence to a common baseline? How would GDP reductions be allocated and agreed across countries? It would, in fact, be extremely difficult to achieve global coordination on a degrowth strategy.

Also, in the rich countries, debate over degrowth is a luxury that can be afforded given their existing high living standards whereas, in the impoverished nations, the priority is addressing basic needs for food, health and education. In emerging economies, people are keener to increase their living standards than to address complex climate change challenges.

Central to the degrowth approach is that it continues to rely, albeit on a reduced scale, on existing technologies, which are not environmentally sustainable. It is important not to lose sight of the fact that the primary goal of climate change efforts is achieving net zero emissions sooner rather than later. This will require innovation and new technologies and, crucially, more extensive and efficient use of existing sustainable technologies to reduce the harm from current production processes. The focus must be on sustainable production and absence of emissions that are harmful to natural capital rather than on reduction or degrowth. To reach net zero will require substantial investment in research, development and infrastructures. Significant investment will be unlikely if the aim is curtailing production.

Making economies more sustainable

Although degrowth is not the ultimate solution to society's problems, there are substantial adaptations needed within the existing framework. We must curtail various production processes that harm the atmosphere and deplete our natural capital. To assume that the current global economic structure will lead effortlessly to net zero is a fallacy and those advocating this likely have ulterior motives to maintain the status quo and its associated benefits.

I want to emphasize two points. Firstly, it is important to curb utilization of environmentally detrimental technologies and transition to greener alternatives that work in harmony with nature's regenerative capacity. The atmosphere has a specific annual CO_2 assimilation capability, and this limit should be our guiding principle for achieving overall CO_2 emissions, again coming back to the concept of net zero emissions. Similarly, our use of natural resources, such as forests, oceans and freshwater, should be governed by their ability to self-replenish. We must engage with these ecosystems in terms, only, of their regeneration ability, to ensure that they remain stable over time.

Second, the focus on energy conservation must be heightened and extend beyond individual eco-conscious behaviours to encompass the broader structure of our economic system. Our current economic model often neglects the significance of prolonging product lifespans and optimizing recycling processes. It is a system that is geared towards consumerism and products with built-in obsolescence that continuously boost manufacturers' profits through increased sales. This planned obsolescence can be seen for example in the planned slowing of smartphone batteries over time and in the cost of spare parts being maintained high enough to make repairs uneconomic. Also, discontinuing production of certain automobile models leads to surges in vehicle parts thefts and makes maintenance of older models and products deliberately difficult.

Contrast this with a system in which products last for longer, resulting in reduced sales volumes, but enhanced sustainability. To change the current paradigm, manufacturers perhaps should be obligated to recycle a certain percentage of their discarded products. This might spur them to prioritize recyclability and enhance product longevity through effective customer service. In many regions, product servicing is not available. Faulty items, even recent ones, are discarded and replaced.

We should perhaps envisage a system where durable goods, such as cars and other appliances, remain under the manufacturers' ownership and consumers lease rather than purchase these items from their manufacturers. Subscription models encourage producers to maximize durability and recyclability, since their profits depend on product longevity and recycling efficiency. This approach, which has been described as part of the "circular economy", could drastically reduce waste, and promote a more sustainable interaction with our environment.

The emphasis on minimizing waste leads to a broader perspective on economic growth. Equating economic growth solely with gross domestic product (GDP) is an oversimplification that we need to address. The GDP measure, developed in the 1930s by economist Simon Kuznets, emerged as a tool to gauge national economic performance and, particularly, to aid recovery following the Great Depression. GDP represents the total monetary value of all the final goods and services produced over a specific period within a country's borders. It includes household consumption, government consumption, public and private investment, and net exports, but does not take into account the informal economy, other non-monetary exchanges, or depletion of natural resources. It also overlooks factors such as income inequality, quality of life, and environmental impact, which has led to criticism of its adequacy as a single measure of national economic success and well-being. Nevertheless, GDP remains the primary indicator used worldwide to compare national economic prowess and rates of growth.

While it gauges annual production of marketable final goods and services – analogous to an individual's yearly income – an important shortcoming of GDP as a metric is that it does not measure the waste generated or the "natural" and "physical" capital depreciation involved in their production. Physical capital refers to tangible assets created by humans and used in producing goods and services. This includes items like machinery, buildings, vehicles and infrastructure. Natural capital provides ecosystem services that benefit humanity, such as clean air and water, fertile soil for agriculture and the raw materials for construction and energy production.

Physical capital is derived from, but is distinct from, natural capital in that it represents the human-made transformations of natural resources into tools that facilitate production. Both forms of capital are fundamental for production: natural capital provides the raw materials and physical capital represents the tools and infrastructure that convert these materials into consumable goods and services. In the case of GDP, no deductions are made for physical capital depreciation or, more crucially, depletion of natural resources. For example, atmospheric degradation due to emissions, and contamination of land and water by waste and pollutants, are not included in GDP, which would seem to underscore the need for a more comprehensive measure.

To illustrate this further, the machinery involved in car production and its wear and tear (depreciation) within a car manufacturing facility, are not considered in the calculation of car production output. The GDP calculation counts the number of cars produced, valued at their market price after deducting the cost of intermediate goods such as the components procured from other manufacturers. It takes no account of machinery depreciation and the potential environmental costs related to industrial waste that might affect our natural resources, leading to biodiversity loss.[8] Although the inclusion and measurement of these costs are difficult, ideally, they should be deducted from the GDP measure in order to provide a more comprehensive and accurate economic picture. Interestingly, GDP includes the expenses related to scrapping old appliances and cars, because this is a market service for which the consumer pays.

The solution proposed by economist Partha Dasgupta is compelling and suggests that our income measure should account for the change in GDP after deducting both physical and natural capital depreciation. We could also consider including human capital depreciation.[9] This shifts the focus from GDP to a more encompassing net domestic product (NDP). When evaluating "growth", it would be more insightful to measure the change in total wealth across a national economy. This would include the cumulative changes in physical, human and natural capital after accounting for their respective depreciation. Dasgupta calls this comprehensive measure "inclusive wealth". Inclusive wealth counts growth when our consumption of produced goods and services is below NDP. In this scenario, our saving – derived from the difference between NDP and consumption – more than offsets depreciation of the various forms of capital and, therefore, is a true accumulation of inclusive wealth.

Clearly, these calculations would require a relative price to be assigned to natural capital in relation to physical and human capital. This relative price would represent a value that would reflect the societal worth of natural capital. It would encompass both the market value associated to utilization of natural capital (e.g., the cost of the timber derived from logging) and the valuation of external factors not captured by the market price (such as the forest's diminished capability to sequester CO_2 and biodiversity losses). This latter is analogous to the social cost of carbon (SCC), which quantifies the economic damage associated with an increase in carbon dioxide (CO_2) emissions of one unit, conventionally one tonne, in a given year. This monetary figure is also a measure of the value of the damages avoided through emission reductions (the benefit of a CO_2 reduction).

This cost includes direct damage, such as reduced agricultural yields and harm to human health and, also, the broader impacts on ecosystems and infrastructure due to changing climate patterns. For example, a high SCC is a sign that emitting carbon has significant negative impacts and would justify implementing policies such as carbon pricing mechanisms and investment in renewable energy to reduce emissions. By assigning a cost to the damage caused by carbon emissions, carbon pricing based on the SCC incentivizes businesses and governments to consider the long-term environmental impact of their actions, which would promote protection and preservation of ecosystems against the adverse effects of climate change.

Assigning a price will be essential to protect ecosystems that tend to be exploited, precisely, because of the lack of a designated price. Take the example of the Amazon rainforest, which plays a crucial role in absorbing CO_2 emissions and acts as a giant carbon sink. The process of photosynthesis allows trees and plants to absorb CO_2 and to use the carbon to grow and release oxygen back into the atmosphere. This process helps to mitigate the greenhouse effect by reducing CO_2 amounts in the atmosphere. The vast area of the Amazon rainforest means it can absorb billions of tonnes of CO_2 annually which helps to stabilize the global climate. However, deforestation and forest degradation by the people who inhabit the Amazon are diminishing its capacity to absorb CO_2 and are contributing to rather than mitigating climate change. The health and preservation of the Amazon rainforest, therefore, are vital for biodiversity and its role in the global carbon cycle and the fight against climate change.

The ongoing deforestation of the Amazon rainforest to make room for cattle ranching, agriculture, logging and mining activities has negative global externalities; it reduces its CO_2 absorption capacity. This cost tends to be overlooked by decision-makers, but it is essential that prices should be assigned to natural capital and ecosystems to capture their intrinsic value for humanity and account for these negative externalities. However, determining these shadow prices[10] is as difficult as deciding who should support the costs of using natural resources. Should it be the countries in which these natural resources are located, or the countries where they are utilized, or the developed nations advocating for their protection?

In summary, although degrowth might not be seen as the primary solution for tackling the climate crisis, it undeniably underlines the urgency for transformative shifts to the status quo. However, we have yet to fully address the pivotal questions posed at the beginning of this chapter: Will the transition to net zero potentially usher in economic growth or introduce constraints and economic downturns? We must address and acknowledge the significant long-term advantages of, and short-term obstacles to, the transition to a net-zero paradigm.

The long-run benefits of the transition

Understanding the economic implications of transitioning to a climate-resilient future is intricate and complex, but there are some persuasive arguments that such a transition would be economically favourable. The advantages are clear if we consider the costs and risks associated with maintaining the status quo, which ignores the need for environmental sustainability. Investment in climate mitigation and adaptation could not only avoid the detrimental impacts of unchecked climate change but also could provide new opportunities for growth and innovation.

First, the economic repercussions of climate-induced events resulting from global warming are profound. Natural disasters inflict immediate damage, but also induce long-term economic downturns that affect both infrastructure and human lives. The economic costs associated with rising global temperatures are uncertain, but are expected to be very significant and certainly higher than the costs related to the transition to net zero (see, e.g., Drouet *et al.* 2022). Adapting to extreme climate events requires dedicated investment in two main areas: preventative measures, such as construction of infrastructure to protect against extreme weather events (e.g., construction of sea walls, building reinforcements) and remedial actions, which include post-disaster relief and reconstruction. Research suggests that, compared to remedial measures or inaction, proactive preventative interventions tend to spur higher growth (Catalano *et al.* 2019). However, the high initial costs of preventative measures, combined with budgetary constraints, have led many nations to prioritize post-disaster interventions. This approach often amplifies the human and financial costs of natural calamities.[11]

Second, a shift towards renewable energies is compelling, due to their inherent sustainability. According to the International Energy Agency, solar energy is the cheapest electricity generation source in human history and its use is being accelerated by climate protection policies. Solar power is a renewable source of energy that eliminates the need to extract, import and process vast quantities of fossil fuels. Cheap energy can promote economic growth by reducing the operating costs in various industries, enhancing productivity and spurring innovation.

The need for renewable energy to sustain economic progress and create new industrial and occupational sectors suggests ways the energy transition would be beneficial, provided it can foster industrial progress toward more productive activities. In an economy powered by affordable solar and wind energies, energy-intensive sectors, such as manufacturing, could see a significant decrease in their costs, increased competitiveness, and expansion potential. This cost advantage would also accelerate technology sector growth, including companies specializing in energy-efficient products and services. In addition, the construction industry would benefit from increased demand for solar panel installations and retrofitting of buildings to improve energy efficiency. Employment in and related to renewable energy technology, such as solar panel technicians and wind turbine engineers would increase alongside more R&D staff working on green technology. This should create a ripple effect and increase employment in ancillary services, including logistics, maintenance and management, as the overall economy adapts to and integrates these sustainable energy solutions.

Moreover, the shift to renewable energy sources will have pronounced distributive implications. Countries that traditionally import oil, including numerous European nations and emerging economies such as China and India, would gain substantially in the long run. The savings on importing energy would bolster these countries' energy autonomy and economic resilience. However, nations with major fossil fuel reserves and extensive processing infrastructures would be faced with the risk of considerable stranded assets and would suffer the economic repercussions of a global momentum towards green energy alternatives.

The transition to green energy would set the global economy on a course of sustainability, with higher incomes compared to a scenario marred by significant warming and frequent natural disasters that stifle productivity and inflict economic damage. Over time, this shift can be expected to generate an abundance of affordable energy, derived directly from the sun, reducing reliance on finite and more costly fossil resources. While traditional industries, such as coal, oil and gas, would contract, those sectors centred around renewable energies would flourish. The benefits of the energy transition go beyond economic gains and would protect the environment.

Conservation efforts to safeguard forests, preserve water resources, mitigate pollution, and protect diverse habitats are integral to this transformation. These measures not only conserve our planet's irreplaceable biodiversity but also enhance the economic benefits of the green transition, particularly in terms of reduced air pollution, and would have significant implications for public health and productivity. However, it must be recognized that while the advantages of transitioning to greener energy are long-term, the costs are often immediate, which leads us to the next set of issues.

The short-term obstacles

While the advantages of the green transition seem evident, its achievement involves numerous obstacles, including the immediate tangible costs involved. The concept of stranded assets is an important example of these transition expenses.

In tackling the urgent environmental problem,

certain assets that are currently operational are likely to become redundant. These stranded assets range from the obvious fossil fuel operations to categories such as internal combustion engines which, although still functional, will be prohibited over time. In addition, energy inefficient buildings will require significant investment if they are to retain their market value. This physical asset redundancy will be paralleled by skills obsolescence. As certain sectors scale down, in line with green objectives, the workforce employed in these sectors will find that some of their skills are irrelevant.

These transition costs are not theoretical; they will be manifested in real financial terms and will affect economies, sectors and households. As the energy sector restructures, there could be a period of climbing energy costs to cover the investment needed to develop the renewables infrastructure and potential inefficiencies during the initial transition phase. These higher energy costs, asset depreciation and skill redundancy will have direct economic impacts. Financial institutions may need to reassess the long-term viability of some of their investments and insurers are likely to see changes to their risk profiles. The role of government will be pivotal for designing policies to mitigate these transition risks, support workforce reskilling and ensure energy affordability. This intricate financial, economic and social web will decide the profound and difficult journey towards a sustainable future.

There is an understandable reluctance to bear the costs of what will be the broader, collective benefits from reduced emissions. While businesses and households will have to manage the immediate financial implications, potentially with some assistance from public coffers, the advantages of reduced emissions and a healthier environment will benefit society at large. As we saw in the Introduction this discrepancy between the private costs and public benefits leads to the classic economic conundrum of the problem of externalities. If individuals or entities are unable to reap the full rewards of their positive actions or, conversely, don't have to bear the brunt of their adverse actions, they tend to be less incentivized to act responsibly. This inherent dissonance encourages free-riding behaviour where economic agents evade sustaining the costs if they think others will step in. In such a situation, no one has a clear incentive to act to reduce emissions. In the context of the green transition, this will increase the challenges.

The temporal scope of the climate issue makes navigation of the complexities of the green transition even more difficult. The problem involves tackling the effects of the historical emissions that have lingered in the atmosphere and which have already caused temperature rises of more than 1°C. It is regrettable that this situation is almost irreversible. Current emissions levels are causing further warming and making the trajectory even steeper. While slashing emissions would decelerate the rate of temperature escalation, the goal of net zero emissions, while ambitious, would merely stabilize temperatures at these elevated levels. In other words, the efforts and investment poured into emissions reductions today must be juxtaposed with the relentless upward global temperature trajectory. Also, all current expenditure and sacrifices will benefit mostly future generations; the benefits from curbing emissions will accrue over time and may not be realized, in full, in our own lifetime. This temporal disconnect between cost and benefit adds an additional layer of complexity to climate action, requiring a long-term perspective that values the well-being of future generations as well as our own.

However, the problems are not confined to either generational equity or the complexities of externalities. Public awareness, or lack thereof, is an additional problem. Despite the global stakes, many are uninformed about, or apathetic towards, the climate crisis. Organizational hurdles further impede progress. The logistical difficulties involved in securing permits for large-scale renewable energy projects are significant. Furthermore, despite a broad consensus about the need for renewable energy, local opposition to such projects is frequent and typical of NIMBYism.[12]

The influence of powerful lobby groups, with vested interests in maintaining the status quo, is similarly unhelpful. These groups often wield considerable power in shaping public opinion and policy directions. Finally, there is no agreement about the best policies and strategies to achieve the green transition. Should the emphasis be on carbon pricing to reduce carbon-intensive activities? Or should the focus be on promoting innovation and fostering nascent sectors and technologies? A comprehensive policy portfolio of appropriate market instruments, regulation and innovation is essential to address the climate challenge. Communicating the scope of such a policy reform will be difficult and will require substantial political will.

Before investigating these intricate challenges in more detail in the succeeding chapters, in the rest of this chapter I want to highlight three immediate conceptual obstacles briefly touched upon above: the long-horizon perspective, the free-rider problem and the heightened uncertainty related to climate issues. I provide a brief discussion of each along with some potential solutions proposed by economists. I believe that some of the inertia in reshaping our economic approach to address climate issues stems from the over-reliance of economists on their traditional toolkit.[13] While these frameworks may have served us well in the past, they may not be appropriate for the distinct complications related to climate change. However, I show also that the field of economics is quickly adapting to this new reality.

Three tricky problems

1. *The long horizon.* Climate change is often viewed as a distant concern, requiring nations to persuade their citizens and businesses about the urgency of immediate spending for future environmental welfare. Convincing stakeholders of this immediacy is difficult. A colleague once mentioned to me that: "Professionals are primarily appraised on their short-term achievements. With the effects of climate action being largely intangible in the near term, the motivation to act diminishes". However, this "distant" threat is looming larger than

most anticipated even in the recent past. Scientific data emphasize the shrinking of the window for effective action. Given current GHG concentrations, permissible future emissions limits are reducing rapidly. The 2021 IPCC Sixth Assessment Report (AR6) has estimated that the remaining carbon budget to limit global warming to 1.5°C is around 250–400 billion tonnes of CO_2 (IPCC 2021). At the current emissions rate of about 35 billion to 40 billion tonnes per year, this budget will be exhausted in 6–11 years. The budget that would allow achievement of below the 2°C threshold is estimated to be 1,150 billion tonnes of CO_2, which would be exhausted in about 25 years. However, there are uncertainties accompanying all these estimates. The uncertainty range for the 1.5°C budget is approximately ±200 billion tonnes of CO_2, and for the 2°C budget, is around ±620 billion tonnes of CO_2. Unfortunately, current trajectories lean towards this grim outcome although, for many, a decade is a long time, particularly when tangible results remain elusive.

2. *Externalities.* Polluters often ignore the societal costs of their GHGs and continue with economic activities that fail to account for the repercussions of heightened temperatures. If emissions of GHGs continue not to be penalized, some organizations will likely continue their harmful practices, unburdened by the societal damage caused. The externality issue induces two reinforcing effects: it promotes pollution due to lack of direct costs and diminishes the motivation to reduce emissions voluntarily. Those taking steps to lower emissions bear the brunt of the costs associated to adapting

their operations and receive no compensation for the societal benefits this provides. This imbalance considerably weakens the incentive to cut emissions.

3. *High uncertainty.* The major uncertainty related to predicting climate change effects is well known. For example, the idea that temperature rise impacts are linear is misleading; abrupt climate shifts may occur. We also cannot precisely foresee the aftermath of temperature rises beyond certain limits, nor can we experiment to verify potential outcomes. Exceeding 1.5°C or 2°C might trigger unpredictable climate events or reach tipping points. However, there are uncertainties, also, surrounding possible technological advancements. Technological breakthroughs could simplify emissions reduction efforts. Initially high mitigation costs might be compensated by the discovery of more efficient technologies. However, relying on technological innovations is risky, and postponing action on emissions reductions, in the hope of the emergence of better technologies, could lead to irreversible atmospheric changes. Thus, there are considerable uncertainties related to the trajectory of climate change and the potential for technological advancements.

Economists relish such challenges, given their proficiency in evaluating intertemporal trade-offs, rectifying externalities and managing risk. Investing resources today for the future is beneficial if it sufficiently amplifies tomorrow's consumption to offset today's forgone consumption. The worth of tomorrow's consumption is deemed appropriate if the utility from an additional unit of consumption tomorrow, discounted to the present, compensates for the cost of deferring today's consumption by the same unit. This depends largely on the chosen intertemporal discount rate, which is the rate used to determine the present value of future rewards. This rate reflects how much we prefer consumption today over consumption at a future date.

Consider the example of planting a tree. An economist would view this as an intertemporal trade-off: the initial cost includes the price of the sapling and the effort involved in planting it. Over time, the tree grows, providing various benefits, such as shade, beauty and, potentially, fruit. If we assume the tree's benefits over the years – when discounted back to present value using an intertemporal discount rate – are greater than the tree's initial cost, the investment is justified. The intertemporal discount rate reflects how much we value immediate costs against future benefits. If we place a high value on future benefits, we might choose a lower discount rate, making the long-term benefits of the tree seem more valuable today, thereby justifying the initial investment.

Similarly, the decision to quit smoking can be assessed through an intertemporal lens. The immediate cost is the effort and potential discomfort of breaking an addiction, but the long-term benefits – improved health, reduced medical expenses and increased life expectancy – are substantial. Economists would suggest that if the present value of future health benefits, discounted by an individual's intertemporal rate, outweighs the present discomfort, then stopping smoking is a rational investment. This decision also involves externalities since secondhand smoke affects others, and rectifying this negative externality provides additional social benefits.

Although individuals might have personal views on the ideal discount rate, determining it for entire societies becomes significantly more complex. Investment in emissions-reducing technologies today mainly serves the very young and future generations, raising the question of how much emphasis we should place on their well-being. The "social discount rate", that is the discount rate that society applies in its decisions,[14] is a topic of discussion among economists. Some attempt to deduce the long-term investor discount rate by analysing transactions related to long-term assets in the financial or real estate sectors. Once consensus on the social discount rate is reached, economists can assess the pros and cons of a long-term investment and determine its economic viability.

Economists also have devised a strategy to address externalities. As mentioned, externalities are costs or benefits arising from an economic activity that affect third parties who are not directly involved in the activity itself. A Pigouvian tax, which is a tax imposed on any market activity that generates negative externalities, can correct market outcomes by being set at a level which induces the polluter to cut back on pollution to the social optimal level. It, thus, aligns private costs to social costs, leading to a more socially efficient market outcome.

An example of a Pigouvian tax is a carbon tax, which is levied on the carbon content of fuels, to account for the costs of climate change associated with carbon emissions. A carbon tax functions as a charge imposed on companies and, in some cases, individuals for using carbon-intensive resources such as coal, oil and gas, which release damaging carbon emissions into the atmosphere. The primary objective of this tax is to increase the cost of these carbon-emitting sources and, thereby, incentivize both companies and individuals to transition to cleaner energy alternatives or introduce innovations to reduce pollution. This transition is pivotal for mitigating environmental degradation and curbing the effects of climate change.

When the true social costs of GHG emissions are factored in through taxation, businesses and individuals are more likely to internalize the repercussions of their activities and adjust their operations accordingly. To illustrate, envisage the implementation of a substantial carbon tax on emissions. This, inevitably, will raise the operating costs of industries with high CO_2 emissions, prompting a contraction in these sectors. Concurrently, it will catalyse growth in other sectors with smaller carbon footprints. During the transitional phase, while some sectors may experience a downturn, the surge in infrastructure investment and R&D, geared to zero-emissions processes, potentially will bolster overall economic activity.

It is important that the tax accurately reflects the social cost of emissions. As previously mentioned, in the US, the estimates of the social cost of emissions are referred to as the social cost of carbon (SCC). This demands meticulous and complex estimation. The difficulty resides in projecting the long-term discounted costs generated by an additional unit of emissions. These costs encompass the implications of rising temperatures on economic activities, the adverse effects of pollution on human health and, in the context of the growing global population, the amplified negative impacts on an even larger number of people.

When determining the social cost of emissions, it is imperative that the overall damage, which will be borne also by future generations, is discounted. This requires estimation of a social discount rate. A precise estimate makes it possible, in theory, to levy a tax on CO_2 emissions to reduce them to the optimal level, that is the level where the marginal cost of reducing an additional tonne of emission is equal to its marginal benefit given by the reduction in the overall future costs of carbon discounted to the present. Although these measures have been calculated and applied in various contexts, it should be noted that both the intertemporal discount rate and the social cost of emissions are figures shrouded with significant uncertainty.

Finally, economists are expert at navigating the complexities of uncertainty. In the context of climate change, these uncertainties can influence asset values significantly. Climate change introduces risks that can sway asset evaluations. The economics discipline has demarcated these risks into two specific categories: "transition risks" pertain chiefly to the unpredictability surrounding government actions. For instance, will government introduce a carbon tax? If so, when? Might government increase its level in the future? "Physical risks" relate to the direct consequences of climate-driven events for assets such as real estate or production facilities. For example, if a particular region suffers increased flooding, how might this influence the valuation of the assets in that region?

Economists often hold the view that financial markets possess the finesse to accurately gauge and price these risks. This means assets with heightened climate-related risks would naturally attract higher costs of financing since investors would seek compensation for accepting the additional risk. Consequently, for equivalent expected returns, capital gravitates towards assets seen as less vulnerable, especially those with lower emissions and which demonstrate resilience to climate-induced events. Given this dynamic, many economists contend that financial mechanisms can allocate resources in accordance with climate risks and, thus, play a pivotal role in the broader battle against climate change.

Collectively, these discussions would seem to imply that economists have constructed a comprehensive framework to address climate change and that leveraging these concepts will provide a solution to the climate crisis. This is not the case. The solutions championed by economists so far have not in fact significantly advanced the debate. The arguments over which we now turn to.

The discount rate and the carbon budget

We have already mentioned that climate science suggests that, given current emissions rates, we have a limited number of years before atmospheric GHG concentrations reach levels that make a global average temperature rise of more than 1.5°C or 2°C nearly certain. Suppose that society chooses a short-term, self-interested vision, with little concern for the well-being of future generations. Were this scenario to prevail, the current GHG emissions reduction rate would be insufficient and, potentially, would mean higher temperatures for many centuries.

Is this the best outcome? Does the economist's method of determining the present generation discount rate (specifically, that related to the segment currently investing in long-term assets) stand up to scrutiny? Should the goal not be universal consensus on environmental preservation and preventing significant temperature rises, even if not the "optimal" choice based on the estimated discount rate of the current generation? Why should the present generation be responsible for deciding the fate of the planet? More crucially, why should today's generation be allowed to make unilateral decisions that could lead to continuous and persistent temperature increases and potentially catastrophic consequences – in the extreme case, annihilation of humanity – that take no account of the well-being and possibly existence of future generations?

In this context, the choice is not one of leaving the problem to future generations to solve; it is about the legacy of a persistently warming planet over several centuries, which could make our planet uninhabitable and wipe out any prospect of a return to former conditions. In effect, this would be privileging our near-term comfort at the expense of the long-term welfare of our descendants. This reveals a profound externality:

imposing costs on future generations who are not represented in today's decisions.

Resistance to carbon pricing

The idea of imposing a Pigouvian (carbon) tax on emissions is robust in theory but is difficult to enact globally. The lower consumption base of developing countries means that they are exposed to a higher marginal utility of consumption. In other words, they prioritize immediate consumption growth. Also, their role in GHG emissions is often comparatively minimal, as discussed in Chapter 1. Therefore, why would they willingly implement an emissions tax that implies a minor impact on future global temperature rises? There is limited incentive. It is evident that the largest emitters should shoulder the most responsibility, given their significant contribution to GHG concentrations. Their historical emissions make the industrialized nations primarily accountable for GHG concentrations. At the same time, major emerging economies, such as China and India, are among today's top emitters and are forecast to remain so in the future.

While a carbon tax might seem economically sensible, for these nations, its associated immediate costs are likely to be a deterrent to its imposition and, especially, as the benefits lie in the distant future. To sum up, although a Pigouvian tax might seem the optimal theoretical solution to address externalities and might appear ideal to a global benevolent overseer, it may be at odds with the practical realities. We shall discuss more on this issue in Chapter 5.

Can finance distribute climate risk effectively?

Both transition risks and physical risks are shaped by government policies and regulation, making them inherently dependent on governments' action. Should governments actively pursue policies to curtail emissions, transition risks would increase, but physical risks might diminish. If governments procrastinate and fail to implement robust policies to lower emissions, rising temperatures become more probable, which increases the physical risks. Currently, the predominant climate change risks pivot around public policy uncertainty and the uncertainty surrounding the choice of transition pathway more broadly. The uncertainty ranges from a business-as-usual approach, which means minimal efforts and significantly rising temperatures, to an approach in line with the Paris Agreement target of temperature increases limited to 1.5°C.

So, what role can the financial sector play in alleviating this uncertainty? In truth, unless financial strategies are aligned with government policies, the potential is limited. Expecting the financial realm to prioritize eco-friendly enterprises in the absence, especially, of cogent public policies to meet net zero objectives, would seem somewhat unrealistic despite what economists might claim. Without such interventionist policies, capital is likely to continue favouring sectors that yield higher returns, as in the recent case of the fossil fuels sector. Government policies must pave the way to enabling finance to support the climate transition.

In essence, the misalignment between the intertemporal discount rate and the finite carbon budget poses a critical dilemma, since current emissions rates threaten to surpass the threshold for safe global temperature rises. The choice is stark: continue prioritizing short-term gains or consider the long-term welfare of the planet and of future generations. Despite the theoretical appeal of a Pigouvian tax, its global implementation would be difficult, especially in the context of developing countries with immediate development priorities. Meanwhile, the finance sector's ability to mitigate climate risk is hampered by the absence of clear government policies. The urgency of these issues sets the stage for Chapter 3, which investigates the policies and actions currently being undertaken by governments, local authorities and the private sector, to reduce GHG emissions and steer private finance towards green initiatives, a cornerstone of efforts to achieve the net zero target.

3

Net zero policies

"We are the first generation to feel the effect of climate change and the last generation who can do something about it".
Barack Obama, remarks at the First Session of COP21.

Governments wield significant power, particularly through their fiscal strategies and regulatory mandates, to achieve their emissions targets and combat the threats posed by climate change. This chapter begins by examining the fiscal tools available to policymakers in supporting the green shift. These tools range from emissions taxes to incentivizing investment in clean sectors such as renewable energy, and improving energy efficiency, and exemplified by inducements to upgrade the energy efficiency of buildings. Governments, also, can make direct public investments in providing sustainable transportation and enforcing regulations to reduce pollution. We investigate the ideal mix of these fiscal tools to not only steer the economy towards net zero but also to respect fiscal prudence (to avoid overburdening the state's finances). This balance is vital, as the growing climate crisis and the rising costs of climate adaptation and mitigation put a strain on public coffers. We shall also discuss the latest governmental initiatives aimed at reducing GHG emissions and closing the gap between pledges and real actions.

We shall continue by interrogating the pivotal role of financial markets. The green transition is an ambitious venture that will require colossal investment and will require much more than government funding alone. The private sector will have to act. Despite a recent noticeable uptick in climate-focused private investment, much more finance will be needed. Public policy has a crucial role in ramping up the appeal of green investments and attracting potential funders and backers. As in the case of carbon pricing – assigning a cost to pollution to level the playing field for the more expensive eco-friendly emitting activities – only government incentives to make investment in renewables more attractive for low-emitting sectors can kickstart entire new industries. At the same time, financial regulators are nudging industry to evaluate how climate change – from policy shifts to natural disasters – might affect their financial stakes. Recognition of these risks is making financing for high-emission activities, and for activities vulnerable to climate impacts, more expensive, which is favouring activities that are both low-emissions producing and resilient to the physical threats of climate change.

The fiscal tools available

Policymakers have various strategies at their disposal to drive the climate transition. It is likely that a blend of these tools will be required to smooth the transition path. It is likely, also, that accepting higher levels of public debt might, paradoxically, support long-term fiscal health, if the additional funds support the green shift. Despite numerous obstacles, public policy remains a cornerstone of net zero planning and this chapter proposes some actionable strategies to enable this critical transition.

Many governments, and particularly governments in developed countries, have pledged to reduce emissions significantly. However, in many cases, as we have seen already, policies and the governance needed to fulfil these promises are insufficient. The integration of climate objectives into financial planning is in its infancy. Green public finance management (GPFM) is similarly at a very early stage, even in the developed economies (Gonguet *et al.* 2021). A survey by the Organisation for Economic Cooperation and Development (OECD) reveals that a staggering 60 per cent of OECD member countries are not engaged in GPFM (OECD 2021). Only a handful of nations are practising green budgeting, primarily for environmental impact assessments to guide fiscal decisions. Europe's approach to green budgeting is patchy and employs a variety of nascent methodologies.[15] Even the frontrunners in this area are only scratching the surface.

Also, the position of the green transition on political priority lists is being continuously eroded by the emergence of unforeseen crises. The 2022 surge in gas prices, spurred by Russia's invasion of Ukraine, encouraged a shift to cheaper, but more polluting, fossil fuels such as oil and coal. Europe has partially missed its chance of leveraging these price hikes to dampen energy consumption. Instead of offering targeted aid only to the most vulnerable people and sectors, governments have intervened broadly, therefore indirectly supporting the demand for energy. These government interventions have eaten into available public funds and have somehow contradicted the imperative to cut emissions.

Economists have long touted carbon taxation or cap-and-trade schemes, such as the EU emissions trading system, as effective for cutting emissions. A cap-and-trade carbon emission system is a market-based approach that aims to reduce greenhouse gas emissions. Under this system, a government sets a limit, or "cap" on the total amount of certain greenhouse gases that can be emitted by industries or sectors covered by the regulation. This cap is usually reduced over time to achieve the overarching goal of reducing emissions. Companies or other entities are given emission permits, or "allowances", and each allowance permits the holder to emit a specified amount of emissions. These allowances can be bought and sold on the market, hence the "trade" part of the name. Companies that reduce their emissions more efficiently can sell their extra allowances to others that might find it more expensive or challenging to reduce their emissions. This creates a financial incentive for companies to lower emissions because they can profit from selling their excess allowances. The flexibility of buying and selling allowances in this system promotes cost-effective emissions reduction across the market.

These fiscal mechanisms assume that emissions carry a societal cost and, thus, companies should cover the costs of their GHG emissions. This idea is in line with the concept of externalities: polluters create a negative impact that a Pigouvian tax aims to rectify. Despite being a pillar of climate-related fiscal policy, the implementation of carbon taxes globally is underwhelming. A mere 20 per cent of global emissions are affected by carbon pricing initiatives, with the average price around a few dollars per tonne. This contrasts sharply with the $75 per tonne estimated by the IMF for effective emissions reduction and maintenance of global warming at below the critical 2°C threshold (Parry 2021). In addition, the presence of fossil fuel subsidies in several countries is acting as an inverse carbon tax and further undermining the fight against climate change.

Recently, a more nuanced stance on CO_2 pricing has emerged globally. If green energy capacity is inadequate or if green technology is not seamlessly integrated with existing energy systems, there is an acknowledgment that even substantial CO_2 and energy cost rises might not provide a huge inducement to shift to lower-emissions production methods. This will result in firms and households struggling to switch from high emissions to green energy sources, despite potential price incentives. The 2022 energy crunch in Europe is an unavoidable reminder: in the absence of sufficient green energy options to offset gas use, even soaring prices might not induce a significant switch.

In the short run substitutes are limited and, potentially, dampen the effectiveness of CO_2 cost hikes. For instance, simple imposition of a carbon tax or increasing the carbon tax may do little to diminish the use of diesel/petrol-driven vehicles or boost electric vehicle adoption, especially in the absence of an efficient charging infrastructure. Direct investment in the required infrastructure might have a greater impact on increasing take-up of electric vehicles than taxing CO_2 (Stock & Stuart 2021). Also, carbon taxes in environments with scant green alternatives are likely to increase energy expenses disproportionately, and increase the burden on low-income households, whose energy costs usually account for a large share of their income. This reality creates a significant obstacle for policies aimed at levying taxes on emissions.

These considerations are leading policymakers to contemplate a more comprehensive public policy mix that not only raises emissions costs but also simultaneously drives the development of more robust low-emission energy production financed by public spending and the introduction of tax incentives and subsidies for private investment. Fiscal policy offers a suite of tools that include private investment incentives, public mitigation projects and direct regulation. These diverse measures are aimed at phasing out high-emissions activities while bolstering low-emissions energy production capacity.

Next, we explore how leveraging a combination of fiscal incentives and direct public investment inevitably will increase public debt. Addressing the redistributive effects of the transition, which is an inevitable part of the process, will also add to the pressure on public budgets. The success of such fiscal strategies will hinge on their credibility; that it is not enough for policies to be well designed and well targeted, they must also be perceived as enduring rather than temporary fixes. Fiscal sustainability is a prerequisite for fiscal strategy credibility and refers to a government's ability to sustain current spending, tax, and other fiscal policies over the long run, without risking insolvency or defaulting on some of its liabilities or promised expenditures. Currently, uncertainty regarding the long-term soundness and sustainability of public initiatives to promote the climate transition is a significant cause of hesitation from the manufacturing sector and global finance to commit to emissions reductions. This concern has been highlighted by the G30 Working Group (2020).[16]

The fiscal costs of the green transition

Public budgets are influenced appreciably by three key climate-related interventions. First, measures designed to support climate "mitigation" efforts, which aim to slash emissions; second, "adaptation" strategies geared towards managing and mitigating the costs associated with extreme climate events, both pre-emptively and after the fact; and third, measures to manage the distributional impacts of the green transition. This last intervention includes providing support to workers in industries that are contracting or being phased out due to the shift away from fossil fuels and providing aid to households and businesses likely to have to bear the higher energy costs during the transition, and the overall inflation likely to be triggered by an increase in energy prices, as happened in the recent cost-of-living crisis.

In the context of mitigation measures, there is a strong case for proactive fiscal policies, such as subsidies and direct public spending, to complement emissions pricing and regulation aimed at emissions reduction. First, green energy production must increase, which will likely require a mix of public and incentivized private investments. Second, a vigorous fiscal approach will be needed to buoy the economy during the transition to cleaner energy and avoid high emissions costs from causing a spike in energy prices and depressed production.

By focusing fiscal efforts on green sector growth, we can mitigate emissions and the broader effects of climate change, providing the dual benefit of immediate economic stimulation and long-term environmental gains. This strategy should spur economic activity in both the short and longer term, potentially helping to stabilize, and eventually even reduce, the debt-to-GDP ratio. That is, proactive fiscal policy will be crucial for bolstering green investment and countering the negative impacts of increasing emissions costs on high-emissions industries.

These costs will depress activity in these sectors, whereas fiscal support and investments in sustainable industries will promote low-emissions production. Therefore, by curbing pollution-heavy sectors and fostering growth in cleaner sectors, fiscal policy will play a decisive role in expediting the green transition. Integrated economic and climate models, that take account of reinvestment of carbon tax revenue in green tax credits and public projects, suggest that such approaches are not only environmentally sound but also fiscally prudent over the long term. While, initially, public debt relative to GDP may rise due to these green subsidies and investments, over the medium to long term it can be expected to decline, resulting in a more sustainable debt landscape (see Catalano *et al.* 2021).[17]

The costs of adaptations to climate change are difficult to quantify since they depend on forecasting the progression of climate-related adverse events. A 2020 study examined the US, where climate change is expected to affect a range of government financial channels: from spending on services such as healthcare, to transfers such as income support, to tax revenue and specific adaptation investments such as construction of sea defences. This study suggests that, in a high emissions scenario, climate change could raise overall public consumption and transfer needs by approximately 1.45 per cent by 2050, with a significant portion of this increase driven by healthcare costs (Barrage 2020). It should be noted that this figure is likely to understate the overall costs, given the limited government intervention in environmental catastrophes in the US and the reliance in that country on private insurance to cover many costs, particularly those associated with property and infrastructure damage.

Obviously, different countries will face different climate challenges. Some will be required to manage ever more severe climate events. The recent debate over possible compensations for "loss and damage" and recovery efforts in response to the harm caused by climate events in vulnerable countries, especially those contributing the least to GHG emissions, is testament to these differences. This debate is especially salient in the context of the agricultural sector and, particularly, smallholder farmers in developing countries where climate change is deepening existing inequalities by disproportionately affecting those with fewer resources and less capacity to adapt.

Finally, the green transition will require significant actions to mitigate emissions, whose distributional impacts are forecast to be profound. Certain industries may face significant downsizing, prompting the need for compensation for affected workers. These considerations will be critical to prevent the emergence of political resistance to environmental reforms. Opposition groups are likely to emerge and, potentially, might be able to influence negotiation of substantial concessions. While the fiscal cost of such compensation is difficult to forecast, its potential magnitude should not be underestimated.

These points raise questions about whether the public expenditures planned by governments worldwide will be adequate to meet their climate goals. As discussed in Chapter 1, the evidence is unequivocal in suggesting that current funding levels are insufficient. While numerous nations are implementing or contemplating measures to support mitigation and adaptation, the challenge remains daunting. The EU, a pioneer of decarbonization efforts, provides a case study of the complexities involved. A European Environment Agency analysis indicates that, to meet 2030 objectives, all major eurozone countries will need to reduce their emissions by between 2–8 per cent annually – a much more rapid pace than has been achieved historically.[18] The US has escalated its decarbonization initiatives with the passing of the Inflation Reduction Act and the Infrastructure Bill, and China's decarbonization agenda is developing, although the specifics are rather opaque. In the next section we explore developments in the three main economic regions which encompass almost half of global GHG emissions.

What measures are governments implementing?

The European Union

The EU is probably the most advanced region in terms of legislation and regulation to reduce emissions and protect the environment. In particular, the EU has approved a climate law and its climate agenda includes a range of measures. The EU's overarching strategy for tackling climate change is encapsulated in the European Green Deal, a comprehensive framework introduced by the European Commission.[19] This ambitious blueprint provides a roadmap for the EU's transition to a modern, resource-efficient, and competitive economic area with net zero GHG emissions by 2050, and the decoupling of economic growth from resource use. The European Green Deal aims explicitly at a just and inclusive transition that ensures the inclusion of all individuals and regions as the EU moves towards a sustainable future.

The "Fit for 55" package of measures falls under this umbrella and represents the EU's commitment to reducing net GHG emissions by at least 55 per cent by 2030 compared to 1990 levels. It is believed that these measures would make the 2050 climate neutrality goal realistic. It is a comprehensive package of proposals, which include revisions and updates to existing legislation laws across a range of sectors that include energy, transportation and infrastructure. The EU's emissions trading system (ETS) has been extended recently to more sectors, with a decreased cap on emissions. The extended scheme includes emissions from buildings, road transport, and heating, not previously included in the ETS. In parallel, new initiatives have been implemented to boost renewable energy use, improve energy efficiency, and impose stricter emissions standards on vehicles.

Among these initiatives are the recently published directives on automobiles and green homes. The first represents a transformative shift in the automotive industry. The directive mandates an end to sales of new CO_2-emitting cars by 2035 and is a significant step towards achievement of Europe's ambitious climate goals. The automotive directive requires that all new cars sold from 2035 onwards must be zero-emissions vehicles; it includes an intermediate target date of 2030 when there should be a 55 per cent reduction in CO_2 emissions from cars, compared to 2021 levels. The directive exempts cars that use e-fuels, which are produced using captured CO_2 and, thus, are categorized as carbon neutral.[20] This exemption was negotiated by Germany. Although the e-fuel technology is not widely available, this exemption legalizes the sale of new e-fuel-only cars post-2035. It has been met with mixed reactions from EU member states and car manufacturers, with some considering use of e-fuel as allowing continuation of the traditional internal combustion engine and some being sceptical about the cost-effectiveness and widespread use of e-fuel.

To boost buildings' energy performance, the EU established a legislative framework that includes the Energy Performance of Buildings Directive (EPBD) 2010/31/EU and the Energy Efficiency Directive 2012/27/EU. Together, these directives promote policies that will help: (1) achieve a highly energy efficient and decarbonized building stock by 2050; (2) create a stable environment for investment decisions; and (3) enable consumers and businesses to make more informed choices to save energy and money. Both directives were revised in 2018 and 2019, as part of the "Clean Energy for All Europeans" package. In 2020, the Renovation Wave strategy, as part of the European Green Deal, set the objective of at least doubling the annual rate of buildings' renovation by 2030. At the end of 2021, the European Commission adopted a major revision (recast) of the EPBD (the so-called "Green buildings" directive), in the "Fit for 55" package. The "Green buildings" directive is an important component of the package: it aims at accelerating building renovation rates, reduce GHG emissions and energy consumption, and promote the uptake of renewable energy in buildings, by focusing on the worst performing 15 per cent of EU buildings. Although this upgrading and renovation work will be costly, it is seen as an investment that will result in significant energy cost savings over the long term. The EU is providing substantial economic support and individual member states will be expected to produce national building renovation plans that include funding and support mechanisms.

The European Green Deal framework allows individual member states to publish customized national policies that will contribute to the collective emissions reduction targets.[21] The agreement between the European Council and the European Parliament on reform of the EU ETS and creation of the Social Climate Fund to provide financial aid to help citizens finance investments in energy efficiency, new heating and cooling systems and cleaner mobility, represents significant progress. This agreement also includes the ambitious target of reducing GHG emissions in ETS-covered sectors by 62 per cent by 2030 relative to 1990 levels.

As the EU continues to implement and refine its climate policy, these actions represent critical steps towards a resilient and sustainable future and clear acknowledgment of the need for social equity and economic viability in the transition to a green economy. These are important policy measures and ambitious emissions targets but are still not enough to put the EU economy firmly on the path to net zero by 2050.

The United States

In the US, the Inflation Reduction Act (IRA), signed into law by President Biden in 2022, marks a historic investment in the country's commitment to climate action. Its goals are ambitious, and it should help to achieve energy security and stimulate economic growth through job creation and reduced costs for households, based on containing energy costs and subsidizing energy-saving property improvements. This sweeping legislation involves the injection of funds into a wide range of sectors and formulation of strategies to pivot America towards a sustainable, green future. Central to the IRA agenda is support for clean energy production. By injecting capital into renewable energy projects and providing more clean electricity resources, the IRA sets the stage for a profound shift from traditional fossil fuels to sustainable alternatives. The focus is not just on cleaner energy generation; it also includes manufacturing. The aim is to cultivate a vibrant domestic market for clean energy technologies, to create a self-sustaining ecosystem of innovation and production within US borders.

On the consumer side, the IRA addresses energy-saving property improvements and rooftop solar installations, in a bid to reduce utility bills and democratize energy generation. These measures are anticipated to ripple across the economy, fostering job creation in new clean energy sectors and breathing new life into existing manufacturing industries. It will require significant energy grid upgrading and envisages an infusion of investment into domestic manufacturing, particularly in clean energy technologies. The IRA instruments are varied and range from tax incentives designed to lower the cost barrier to adoption of clean energy, to funding programmes to catalyse the broader transition to a clean energy economy. It should be noted that the benefits provided by the IRA are subject to conditions to ensure that the funding and support translates into high-quality jobs, fair wages and apprenticeship opportunities.

However, the IRA falls short in that it does not impose a federal carbon tax. This means that polluters will not have to bear CO_2 emissions costs, which could be a missed opportunity to further drive down emissions by making pollution more expensive for emitters. The IRA also allows new fossil fuel developments, involving the auctioning of oil and gas leases in conjunction with renewable energy projects on federal land. Support for the IRA from Joe Manchin (a Democratic Senator representing West Virginia, the second-largest coal producer state in the US) was tied to speeding up the energy projects permissions process, which, potentially, could benefit fossil fuel projects and has raised concerns among environmental groups about impacts on communities and wildlife.

In terms of social equity, the IRA offers bonus credits for projects that comply with specific environmental justice criteria, in order to ensure that benefits provided by the IRA permeate to all of society, including historically marginalized communities. The rollout of the IRA is showing early promise, with significant investment announced and jobs created within the space of months. A year into its enactment, more than \$110 billion has been announced for clean energy manufacturing investments, resulting in over 170,000 clean energy jobs (The White House 2023).

The IRA could significantly reduce GHG emissions, potentially by 1 billion tonnes by 2030, in line with federal climate goals. Investments driven by the Act are boosting community resilience, with billions of dollars allocated to combat the effects of climate change, ranging from infrastructure improvements to promoting environmental justice. The IRA is resulting in reduced energy costs for families, based on savings enabled by energy-efficient measures and tax credits for clean energy adoption. These improvements are a sign of the transformative shift towards a sustainable and equitable clean energy future.

In summary, the IRA is an important step towards a cleaner, more resilient economy. It encapsulates a broad vision of climate responsibility, economic revitalization and social fairness. Its early successes hint at a future where climate action and economic prosperity, rather than being in opposition, can become twin pillars supporting a robust and equitable economy. Nevertheless, the IRA is not sufficient to achieve net zero emissions in the US by 2050.

China

China's climate policies have come under increasing scrutiny. China is faced with aligning its growing demand for energy with the need for decarbonization. Projections suggest that China's emissions will peak by

2025 – earlier than the 2030 target.[22] However, it has no significant emissions reductions plans for the period before 2030, which is raising concern about the effectiveness of China's climate actions during this critical decade. China's government continues to support fossil fuels for energy stability, despite a shift from limiting energy consumption to limiting carbon emissions.

China's energy transition investments, including non-fossil energy and electrified transport, are significant and exceed those of the next ten leading countries combined. Renewables are expected to play a larger role in meeting energy demands, with China aiming for a 20 per cent renewables share in consumption and a 39 per cent share in generation by 2025. The 14th Five-Year Plan indicates that half of the increase in power demand since 2020 should be met by renewable sources. In terms of policy implementation, the government has aligned the carbon emissions peaking timeline of high-emitting industry sectors with China's 2030 NDC target. Efforts are being made, also, to expand the scope of China's ETS to include high emitting sectors such as cement, steel and aluminium. In addition, China's New Electric Vehicle (NEV) Industry Development Plan is aimed at NEVs accounting for a 20 per cent share of the automotive market by 2025, a target already surpassed in 2022.

China is also focusing on buildings, having set energy consumption caps and is improving energy efficiency in buildings, based on renovation and energy efficiency improvement targets for new public and residential buildings set out in its 14th Five-Year-Plan for Building Energy Conservation. It is also increasing its forest and grassland areas to provide carbon sinks and contribute to achievement of its climate targets. This multifaceted approach to tackling climate change includes regulation, market mechanisms and sector-specific targets designed to steer China towards its 2060 carbon neutrality goal.

However, the pace of decarbonization and China's continued reliance on fossil fuels are problematic. In 2022, China approved the construction of new coal power plants at a pace equivalent to two large plants per week, marking a significant acceleration and the highest level of new permits since 2015. In 2022, China began construction of 50 gigawatt (GW) of coal power capacity, a more than 50 per cent increase on the previous year. In total, 106 GW of new coal power plants were approved in 2022. It must be noted that the actual additional capacity to the grid has remained relatively steady at about 26.8 GW in 2022, aligning with the lowest annual additions since 2003. However, these developments make it unlikely that China will achieve its carbon neutrality goal by 2060.

The need to increase ambitions

Governments and their budgetary strategies are vital for achieving the energy transition. The challenge is enormous; governments must increase the exploitation of fiscal tools, which is likely to increase public debt; they must overcome the many political and socio-economic hurdles to the transition and overcome resistance to change; and they must ensure policy consistency and credibility, in order to shape expectations and actions in the private sector. All of these aspects highlight the need to increase the scope and ambition of climate-related fiscal policies significantly.

Addressing the anticipated rise in public debt is a primary concern for fiscal policy in the context of the green transition. As previously discussed, short-term increases in debt may be necessary to fund green public investments and incentives. While this could cause an initial increase in the public debt-to-GDP ratio, in contrast to a scenario with smaller green investment, it is expected to contribute to its stabilization over the long term. This is because eco-friendly spending is crucial for curbing emissions and mitigating temperature rises and extreme weather events harmful to the economy. Nonetheless, managing the upsurge in public debt, especially considering its current high levels, will be a formidable task.

The second significant concern is political resistance stemming from the inflationary and redistributive consequences of increasing emissions costs. To reach the targets set by the Paris Agreement, it will be essential to implement additional measures that will add to the cost of emissions. Depending solely on expenditure – public investments and incentives – will not work to reduce emissions to a sufficient level and could threaten fiscal sustainability (Catalano & Forni 2021). While green public investments are crucial for facilitating the supply-side transition, they will increase overall domestic demand, which may not necessarily be channelled towards green consumption. Consequently, emission taxation will become an indispensable tool, necessitating corresponding fiscal support for households and businesses most impacted by its effects.

Thirdly, the effectiveness of fiscal policy hinges on establishing measures and standards that set long-term expectations. The continuation of climate policies across different political administrations will be vital for creating a consistent and predictable regulatory environment. Policies that not only have long-term impacts but are also aligned with climate objectives and are fiscally viable are more credible. It must be understood that emissions reduction targets cannot be achieved only through provision of budgetary incentives. A durable and credible decarbonization strategy cannot depend on fiscal incentives alone; it requires a comprehensive and fiscally sustainable policy framework.[23]

There is a real risk of substantial public expenditure in the absence of a coherent and feasible strategy. A credible strategy is crucial for a sustainable transition and achievement of emissions reduction targets. A successful transition will require precise and trustworthy policies to steer the private sector's long-term planning and to channel investment into low-emissions sectors (Group 30 2020). However, policymakers will be competing with other priorities, such as defence, and with the uncertainties of climate and technological advancements. This could lead to poor policy decisions and excessive investment in technologies that might become obsolete. Governments should focus not on trying to pick winners in the technology race, but on creating a framework that encourages adoption of low-emissions technologies and empowering the private sector to make the investment choices. This will require the structuring of taxes and incentives around emissions levels, rather than targeting specific technologies or industries.

Climate goals and financial regulation

The achievement of national climate goals will depend, crucially, on government support including via financial regulation. The amount of investment required for a global transition to net zero is projected to be several trillion US dollars annually. However, many countries make no reference to finance in their NDCs and lack precise estimates of the financing that will be required to achieve the green transition. According to the World Resources Institute, total investment for the 51 countries that include an explicit figure in their NDCs is some $1.5 trillion per year (about 1.5 per cent of global GDP). A recent Climate Policy Initiative report notes that, despite the doubling of public and private climate funding in the period 2011–20, which amounts to $850–940 billion annually, a sevenfold increase will be needed by the end of the current decade to satisfy the demand for investment.

There is an apparent consensus that the private sector will have to bear most of this investment burden related to reducing emissions and to implementing adaptations to accommodate higher temperatures and other climate changes. Since public finances have been stretched by the Covid-19 pandemic, it is unlikely that governments will be able to meet the demand for the large amounts of investment required to achieve net zero.

Investing in green technologies and green projects could provide important commercial opportunities for the financial sector but will be also accompanied by risks. Financial institutions are reluctant to assume more risk at a time when they are suffering the financial impacts of ongoing climate change. Some regulatory jurisdictions are requiring banks and other financial entities to disclose their exposure to climate risks, allowing the market to price these risks appropriately and, thus, render green assets comparatively attractive. However, as global warming escalates the risks to financial assets, and as regulators mandate recognition and pricing of these risks, the financial sector will find it difficult to take on any more risk associated with funding the extensive additional infrastructure projects required for net zero. Increased investment to fund the green transition faces several problems. Private financing will undoubtedly be deterred by the risk inherent in green projects and the additional risk posed by climate change on current assets.

Green investment risks

The primary risk is that green investment involves nascent technologies, whose success is unproven, or which might be overtaken by competing innovations. For example, green hydrogen, produced from renewables such as solar and wind power, is not yet commercially viable and relies heavily on ongoing research and innovation. However, it is seen as an important potential future energy source, as fuel for heat and transportation or as a raw material in chemical and industrial processes. However, its potential as a mainstream technology is uncertain. In addition, in the current global geopolitical landscape, energy independence has become a strategic imperative for many nations. This could lead to some countries pursuing technologies such as green hydrogen despite potentially higher costs compared to other energy sources. Added to the technological and geopolitical uncertainties, there are risks generally associated with large infrastructure projects and, especially, in developing or emerging economy contexts. These risks include political instability, macroeconomic fluctuations, exchange rate volatility, etc.

To manage these risks and other issues effectively might require implementation of public–private partnerships. Many green transition investments, particularly infrastructure projects, are associated with higher risks, or low risk-adjusted returns. Large-scale renewable energy projects, such as solar and wind farms, require significant areas of land. For instance, for safe and efficient operation, a single wind turbine requires a considerable buffer area and solar farms can account for thousands of acres of land area.

Energy production is a strategic sector, critical for national security and economic stability, and green energy projects are frequently integrated into the national grid infrastructure. Grid infrastructures are complex systems enabling the transfer of electricity from producers to consumers. Adding solar power to the grid may require upgrades to transmission lines and increased capacity for energy storage to accommodate the variable nature of solar and wind energy. Finally, many green investments are targeted at emerging markets where demand for energy is growing rapidly due to increasing industrialization and urbanization. For instance, India and Vietnam are investing heavily in renewable energy to meet their accelerating energy needs while trying to minimize carbon emissions. Thus, a synergy between public initiatives and private enterprise will be essential for a successful transition.

This situation is resulting in major private financial institutions calling on governments and multilateral development banks (MDBs), such as the World Bank, to mitigate the risks of green projects to make investment more attractive by improving the risk-adjusted returns. De-risking typically involves public entities offering to guarantee loans for investing companies, based on national or MDB guarantees or equity stakes. These guarantees enable companies to secure financing at more favourable rates, which increases the financial viability of their project. In the case of highly indebted developing countries, the main concern is related to increasing capital financing, such as foreign direct investment, to avoid increasing these countries' external debt.

Alternatively, governments could agree a predetermined price for the purchase of energy produced by the new facility. For instance, under a power purchase agreement (PPA) or a feed-in tariff (FIT) arrangement, a government can promise to purchase the green electricity produced, at a fixed price over a certain period. Such arrangements, by providing a secure and predictable revenue stream, incentivize development of new renewable energy facilities and make the investment less risky and more attractive to investors. Under these types of agreements, the electricity purchased by government may be fed into the national grid and distributed to both residential and commercial consumers. They guarantee that the renewable energy produced will have a buyer and be priced such that investors can recover their costs and make a profit.

For example, a government might agree to purchase solar- or wind-generated electricity at a fixed kWh rate over a period of 20 years. This rate could be higher than the current market price for electricity in order to compensate for the higher initial costs of renewable energy technology and to encourage development of green energy projects, although the cost of producing electricity with wind and solar technologies is already at competitive levels with production costs using fossil fuels. Overall, due to the strategic and infrastructural significance of energy production and distribution, collaboration between the public and private sectors will be essential and will require innovative financing models.

It is clear that to achieve net zero will require the transformation of financing structures. Once again, climate finance was a central focus at the 2023 COP28 meeting in Dubai, and resulted in pledges to the Green Climate Fund, the Least Developed Countries Fund and the Adaptation Fund, although the amounts promised are short of what is needed for a global clean energy transition. The COP28 discussions also debated the need to set a new collective quantified goal on climate finance by 2024, starting from a baseline of $100 billion per year. The need to reform financial systems and establish innovative finance sources was emphasized, but no firm decisions were taken.

The increasing climate risks to financial institutions' assets

The management and mitigation of climate-related risks related to banks and other financial institutions are required to maintain financial stability. Earlier, we discussed two categories of climate risk. "Transition" risks related to policy uncertainties, uncertainty about emissions-reducing regulation and shifts in consumer preferences towards more eco-friendly products, which all affect corporate strategies. "Physical" risks refer to extreme weather events, such as floods, tornadoes and droughts, which can inflict considerable costs on both production and real estate assets. Increasingly, financial regulators are insisting that financial institutions evaluate and integrate these risks into their risk management practices. In early 2022, the European Central Bank carried out a climate stress test on all major European banks. This highlights the need for banks to assess and account for both transition and physical climate risks.

Financial institutions have two main strategies for mitigating climate risks. One involves adjusting their investment portfolios by decreasing funding to the highest-emitting companies, such as those in the hydro-carbon sector, and those companies most vulnerable to physical climate risks, while at the same time increasing their investments in low-emissions companies and those less susceptible to climate-related hazards. The other approach involves providing active support to all types of companies to become more environmentally sustainable and more resilient to climate change. This would require investment to transform their production technologies to lower-emissions technologies and provide funding for adverse climate events.

For example, there is broad agreement that, for two main reasons, the hydrocarbon sector must be maintained, at least in the immediate short term. First, ongoing demand for energy during the transition period must be satisfied, which will involve the hydrocarbon sector companies. An abrupt decline in traditional energy production, with no corresponding increase in renewable options in a context of rising demand for energy due to population growth and desire for higher living standards, is likely to drive up energy prices. This could lead to windfall profits for hydrocarbon producers and create additional costs for infrastructural investments in green energy projects.

The energy crisis in Europe following Russia's invasion of Ukraine shows that higher gas prices – although they should make renewable energy more profitable – in the short term can also slow green investments by triggering a substitution from the expensive gas energy source to cheaper high-emissions sources such as coal and oil.[24] Second, hydrocarbons will remain part of the energy mix for activities such as aviation where currently there are no low-emissions alternatives. The net zero targets implicitly include provision for negative emissions to counterbalance the hopefully small, but persistent, use of hydrocarbons. This means that, for the foreseeable future, a certain level of residual financial backing will be required for the hydrocarbon sector.

Thus, regulators must navigate between guiding financial institutions to reduce their portfolio exposure to climate risks and ensure transparency in relation to these risks, and not impeding the financing of green projects which might carry additional risks. Regulatory policy needs to implement a "brown penalty factor", which essentially means assigning higher risk to assets vulnerable to climate risks. Simultaneously, assets with lower climate risk exposure should benefit from a "green supporting factor", recognizing their reduced risk profile. However, the increase in the cost of risk needs careful consideration. Over-penalizing the financing of traditional energy sectors could reduce their funding prematurely, leading to energy scarcity and high prices. For instance, if regulators mandate additional capital based solely on the borrower's emissions level or physical risk exposure, this might take essential funding away from sectors vital for the transition or impose such severe capital requirements that they will limit the ability of financial institutions to fund new green initiatives. A difficult balance must be struck.

The private sector alone cannot solve the problem

Despite the challenges posed by climate risks, leading international banks recognize the potential profitability of green transition investments, which can be compared to the funding of grand infrastructure projects, such as the railways in the early nineteenth century. However, the risks associated with climate projects are significant. Financial institutions will have to both support high-emissions companies to transition to greener operations and channel substantial investments into expanding the renewables sector and funding pioneering technologies. This will require a willingness to increase the stakes in sectors and projects that carry greater risk and uncertainty than traditionally supported by the banks. The financial commitment required for these investments is substantial and current funding levels are insufficient to achieve the transition targets.

Despite the potential for scaling up private sector financing for the green transition, such efforts will fall short without supportive public policies – a point already established in Chapter 2. The substantial profits enjoyed by oil companies illustrate this dilemma, as they attract investors seeking high returns and fund managers whose fiduciary duties force them to prioritize risk-adjusted returns for their clients. Thus, public policies will play a vital role. Strategies like carbon pricing, promotion of renewable energy through financial incentives, and the establishment of new regulatory frameworks are critical to diminishing the competitive edge that fossil fuels have due to their emission exter-nalities. These measures are essential to alter the eco-nomic balance, making sustainable investments more economically appealing than their fossil-based coun-terparts. To be credible, these policies must be aligned with and reinforce countries' emissions targets. Lack of alignment will render targets a source of uncertainty, rather than a guide, for private sector decision-making.

This chapter has argued for a robust application of emissions pricing and regulation to suppress output in high-emission sectors and for tax incentives/subsi-dies and green public investments to catalyse a surge in green capital. Given the short timeline for achieving net zero and preventing significant temperature rises, prompt and decisive fiscal policy measures are essential. Countries should enshrine climate targets into national legislation, as some have already done through climate laws, outlining mandatory criteria for government pol-icies. Independent entities like climate councils should oversee the attainment of legislated targets. If climate objectives are embedded into law, these councils must be endowed not only with the necessary expertise to conduct proper evaluations but also with the authority to mandate policy revisions. Nevertheless, incorporat-ing such measures into legislation is merely an initial step; we must call for more courageous and accelerated government initiatives.

Quantitative targets should be supported by quali-tative principles. For instance, the "Do No Significant Harm" (DNSH) principle should be used to guide policy. The principle is integrated within the EU's sustainable finance framework. It mandates that any project or financial investment should not cause significant harm to any of the six environmental objectives defined by the EU. These objectives include climate change mitigation and adaptation, sustainable use and protection of water and marine resources, transition to a circular economy, pollution prevention and control, and the protection and restoration of biodiversity and ecosystems. The DNSH principle is a key component of ensuring that activities contributing to one objective do not adversely impact any of the others, thereby supporting a holistic approach to environmental sustainability and resilience. This principle is especially relevant when it comes to managing and allocating funds, such as those within the EU's recovery and resilience facility. The implementation of DNSH involves detailed technical criteria and is a fundamental part of Europe's broader environmental policy and action, especially in the context of the European Green Deal. For example, application of the DNSH principle would have been useful in limiting the European governments widespread measures to limit energy price hikes in 2022 and 2023.[25]

Financial regulators need to be supported by robust public policy in order to apply appropriate financial rules that will drive the green transition. Imposing higher capital requirements on assets with climate risks has several drawbacks in addition to making them costlier to finance. First, potentially it could curtail financial sector lending abilities at a time when significant expansion is required. Second, given the high degree of uncertainty linked to assessment of climate risks, accurate estimation of the costs of financing these assets is difficult and overestimation could unfairly penalize assets otherwise aligned to climate transition goals. At the same time, underestimation would impede the transition. Third and, perhaps, most important, the risk exposure of financial assets to climate change is contingent on the prevailing transition scenario, as discussed in Chapter 2.

In the absence of strong emissions mitigation measures, we will be faced with gradual warming of the atmosphere and increased physical risks while, in the presence of effective government action, transition risks will prevail. Since physical and transition risks affect businesses and financial assets differently, there can be no one-size-fits-all approach to climate risks. Any scheme must take account of the types of government policies implemented to support the transition.

Without timely action, we risk not only the detrimental impacts of climate change but also a scenario in which the state's financial and regulatory involvement in addressing the climate challenge must intensify, possibly in a chaotic acceleration. This could potentially lead us to emergency measures like those enacted during the pandemic, although likely lasting much longer, with profound implications for our democracies. It is crucial that our governments move beyond declarations and toward a well-organized green transition. Yet, even in the case in which advanced countries would implement essential public policies and international finance create the frameworks to support substantial investments, this alone may not suffice. As emissions are now rising most rapidly in large emerging economies like China and India, achieving net zero seems unattainable without a high degree of international coordination. Indeed, coordination among governments is crucial to prevent "carbon leakage", where businesses relocate production to countries with less stringent emissions regulations to circumvent stricter environmental policies.

The EU, for example, is addressing this issue through its carbon border adjustment mechanism (CBAM), which aims to level the playing field by applying a carbon price on imports of certain goods from outside the EU. This ensures that ambitious climate efforts within the EU do not lead to a competitive disadvantage for EU industries due to carbon cost disparities on a global scale. It also encourages producers in other countries to adopt greener practices. Such coordinated approaches are critical to ensuring that climate policies are effective on a global scale and do not simply shift emissions from one country to another. The next chapter will delve further into this international aspect.

4

Are international agreements essential?

"The planet is just too small for these developing countries to repeat the economic growth in the same way that the rich countries have done it in the past. We don't have enough natural resources, we don't have enough atmosphere. Clearly, something has to change."

Mario Molina, interview with *The Chicago Tribune*, 13 December 2007.

Although you might agree wholeheartedly with the arguments and ideas I have presented so far, the international dimension of the climate crisis is likely to induce despondency. Each region on our planet contributes a share to total GHGs, but their repercussions for the climate are global, not regional. For instance, Europe is responsible for less than 10 per cent of worldwide emissions; thus, even if this region were to achieve net zero tomorrow, it would not resolve the climate crisis. The combined emissions produced by China, India and Russia represent around 40 per cent of total world emissions, 75 per cent of which come from China (see IEA 2021). We need to consider the climate strategies of these three nations, from the perspective of contemporary global geopolitics. In the absence of collaboration among and commitment to climate goals by these major economies, is there a chance that we will achieve net zero?

The difficulty involved in securing international agreement on emissions reductions can be illustrated using the famous "prisoner's dilemma". This is a thought experiment, in which two people are accused of a crime. Since there is insufficient evidence to charge one or the other, the police present them with a deal: if neither confesses to the crime, both will receive a short prison term; if only one confesses and betrays the other, the betrayer will go free and the other prisoner will receive a very long sentence; if both accuse one another, they will both receive a long prison sentence. The prisoners are isolated but must make their decision simultaneously. Each prisoner's dilemma is between betraying or not. If one thinks the other will confess and betray, the best response is to confess and accuse the other as well to avoid being the only one accused and having to serve a very long sentence. But even if one thinks the other will cooperate and not confess, the best strategy is to confess (betray the other), as the betrayer will go free. In the prisoner's dilemma, the individual best strategy is always to confess, with the result that both confess and are sentenced to a long prison term, a worse outcome than if neither had confessed. Indeed, it is evident that in this context, the two accused would be better off collaborating and not confessing. The point of this example is that choices which are individually rational can turn out to be collectively self-defeating.

The dynamics of international climate agreements can be considered in this context. In an ideal world, if every nation were to collectively reduce their emissions, this would benefit the entire global community. However, each country might consider it more advantageous to abstain from action and rely on the others to shoulder the burden of emissions reduction. This parallels the prisoner contemplating betrayal for personal gain, even under the expectation that the other will want to cooperate and maintain silence. Consequently, countries become trapped in a cycle of scepticism, with each hesitating to commit to a climate agreement for fear that others will not. The temptation to deviate from collective agreement is especially strong when the individual impact on emissions seems small. This deadlock leads to a paradoxical outcome: no country commits to reducing its emissions to a sufficient level, which is to the detriment of all countries.

The classic free-rider problem is exacerbated in the context of developing and emerging countries as many of these nations contribute only fractional amounts to total global emissions. For example, over half of total world emissions are accounted for by only four countries – China, the US, India and Russia – with two-thirds of world emissions accounted for by the top ten emitters. While this should imply that large emitters might not follow strictly the prescriptions of the prisoner's dilemma, as they are aware that their emissions are enough to warm the planet even if the other countries are reducing them, it leaves all the remaining countries, which tend to be also at the lower end of economic development, prone to free riding. This stance might be exacerbated by the fact that historically these countries have contributed very small amounts to the current GHG levels. For instance, the sub-Saharan African nations, collectively, are responsible for a lower volume of emissions than the US state of Texas. At the same time, they are among the countries most vulnerable to climate impacts.

For these countries to commit to substantial emissions reductions seems illogical and impractical, especially since their national budgets are already over-stretched. Take Bangladesh, whose development challenges are significant; this country would find it difficult to justify prioritizing emissions reductions over improvements to public health and education. In addition, were they to allocate funding to climate mitigation, could this be considered just – to prioritize global environmental concerns over immediate national development needs? This is not a mere rhetorical question; it embodies the ethical considerations at the heart of international climate negotiations in which the distribution of responsibilities and capabilities must be carefully balanced.

In addition to their minimal contribution to atmospheric emissions, the less developed countries disproportionately bear the brunt of the devastation wrought by climate change. In recent years, a series of catastrophic events in sub-Saharan Africa and parts of Asia illustrate this cruel paradox.

In March 2019, Cyclone Idai wrought havoc in Zimbabwe, Malawi and Mozambique, claiming more than a thousand lives and leaving millions starving and lacking basic services. Six weeks later, Cyclone Kenneth became the first recorded tropical cyclone to affect Mozambique. The Horn of Africa has also experienced climate-induced extremes. Rising sea temperatures, a by-product of global warming, have doubled the drought risks in this region with especially severe droughts in 2011, 2017, 2019 and 2022, which decimated crops and livestock, plunging 15 million people in Ethiopia, Kenya and Somalia into a humanitarian emergency due to acute scarcity of food and water.

In Bangladesh, India and Nepal, floods and landslides displaced 12 million people in 2022, exacerbating the devastation wrought by the monsoon in 2020, which, in some areas, was the worst in three decades and resulted in a third of Bangladesh's land area being submerged. Although the monsoon season is typified by heavy rainfall, climatologists have noted particularly intensive precipitation due to warmer sea surface temperatures. The deluge that occurred in Pakistan in 2020 affected 33 million people including 16 million children and was a recent and stark reminder of the region's vulnerability to climate change.

The above events are not isolated incidents; they are manifestations of a changing climate that worsens the effects of climate and causes natural disasters. Their increased frequency and intensity underline the urgent need for global action. Thus, climate change presents a moral imperative: the international community must both reduce its emissions and, also, must support the victims of climate change who contribute the least to the problem. This demands a dual approach: aggressive mitigation to prevent future catastrophes and robust adaptation measures.

Nations that have been, and are being, disproportionately affected by climate disasters are seeking reparation from the wealthier industrialized nations; by virtue of their substantial contributions to atmospheric GHGs, the industrialized countries should bear the financial burden for the consequences of extreme weather events and the toll taken on human lives.

This contention was central to negotiations at COP27 in Sharm el-Sheikh in 2022 and was at the top of the agenda of COP28 in Dubai in 2023. At COP27, the developed countries, for the first time, acknowledged their greater responsibility, although stopping short of pledging specific funding for reparations. COP28 saw several developed nations commit to contributing to a loss-and-damage fund, but the pledges so far are a long way short of the sums required based on existing projections of annual climate-related damage. Despite the apparent merits of the loss-and-damage argument, assigning precise attributions for climate impacts remains a contentious and complex issue. While the industrialized countries might agree in principle to the establishment of a reparation fund, their hesitation to translate this agreement into tangible financial support is palpable. This reluctance is both hampering progress and increasing the divide between the affluent and poorer nations.

Consider, for example, the small island nations in the Pacific, which face existential threats from rising sea levels despite their own negligible contributions to emissions. At the climate summits, these countries have repeatedly highlighted the inequities and called for concrete actions. The continuing debate reflects the broader international climate policy problem of reconciling the developed nations' historical responsibilities with the current and future needs of those most vulnerable to climate change. This debate is constituted by a complex mix of ethics, economics and politics, and high stakes for the communities most exposed to the impacts of climate change.

The crucial issue and extreme difficulties involved in achieving net zero emissions are resulting in a geopolitical landscape characterized by Europe and the US under President Biden on one side, advocating for aggressive climate action, and China's, India's and Russia's more guarded position. This situation is complicating any prospect of unified action. The net zero ambition is appearing increasingly fragile against this backdrop of international tensions.

China's commitment, outlined in its most recent NDCs, for a peak in CO_2 emissions by 2030 and net zero by 2060, sets a significant but distant target. The China–US declaration, announced in the run-up to COP26 in Glasgow, was a notable catalyst for climate action, signalling potential cooperation between the world's two largest emitters. However, invocation of the "common but differentiated responsibilities" principle in China's and India's NDCs underlines that they consider that the obligation to combat climate change should be proportional to each nation's contribution to historical GHG levels and current economic standing. For these emerging economies, recognition of this principle is pivotal to maintenance of their development trajectories and population well-being, and to a refusal to be held responsible for a problem largely not of their making.

The industrialized nations have come first in the development process and disproportionately reduced the amount of atmospheric space for emissions, constraining the room for manoeuvre for the emerging and developing countries coming next in the development process. Emerging and developing nations would then like to catch up, but how can they do so while leaving a smaller carbon footprint? Bridging this divide requires not only commitments but also substantial support from the advanced economies in the form of technology transfer and financial aid to ensure that the race to a sustainable future is fair and equitable.

The emissions trajectory adds to the complexity. China's output of CO_2 is surging rapidly and, even if it will be in line with its NDC pledges, projections suggest a potential global temperature rise of 3–4°C by the end of the twenty-first century – well beyond the Paris Agreement goals. This raises critical questions about the feasibility of current commitments and the urgency for all nations to re-evaluate and reinforce their climate strategies to avoid the most catastrophic impacts of climate change.

There is a clear impasse. Negotiations over the distribution of decarbonization costs are fraught since no country wants to bear a disproportionate share of the burden. However, the advanced economies recognize that they must lead by example and accept that the achievement of net zero by emerging economies will be slower and could result in higher-than-ideal global temperature rises. This leadership is not a merely altruistic action; it is an acknowledgement of historical emissions and current capabilities.

Nevertheless, this leadership responsibility does not absolve the emerging economies from making efforts to mitigate climate change. There is an understanding that all countries must contribute to the global effort, but that this contribution should be consistent with their respective development stages and historical input to the problem. There is a simultaneous underlying assumption that our economies can adjust to moderate temperature increases by investing in resilient infrastructure, developing new agricultural practices, and protecting coastal cities against rising sea levels. While mitigating the worst effects of climate change must be paramount, adaptation strategies are essential to cope with the changes already underway. Lessons about adaptation strategies can be taken from the Netherlands, for example, a pioneer in adaptation to rising sea levels and investment in state-of-the-art flood defence systems. Several nations susceptible to regular drought conditions are exploring innovative water conservation techniques and drought-resistant crops. These examples illustrate the dual approach necessary for climate action: proactive mitigation to curb emissions and strategic adaptation to live with the changes we cannot avoid.

However, it must be emphasized that there are limits to adaptation and significant temperature increases, beyond a certain threshold, can have catastrophic and irreversible effects. While there is an undoubted need for adaptation strategies, they are not a replacement for robust and immediate global emissions reductions.

Struggles in climate diplomacy

At the July 2022 G7 summit in Elmau, Germany, leaders agreed to form a "Climate Club", to increase implementation of the Paris Agreement and drive more ambitious climate action in the industry sector, in particular. The Climate Club recognizes the need to address carbon leakage – the result of relocating carbon-intensive production to regions with less restrictive emissions legislation – and the need to ensure a level playing field in the context of global emissions reductions to meet international standards. The Climate Club's operational framework is still being negotiated and the G7 has enlisted the expertise of major international organizations, including the OECD, the International Monetary Fund (IMF), the World Bank, the International Energy Agency (IEA) and the World Trade Organization (WTO) to provide specific expertise, guidance and support. The leaders agreed that the Climate Club will be inclusive and open to all nations committed to full implementation of the Paris Agreement and its subsequent iterations, including the Glasgow Climate Pact, and the ramping up of climate efforts.

The alliance of nations within the Climate Club can be likened to a relay race team each of whose members is determined to both run as fast as he or she can and to encourage the other members of the team to do the same to ensure the baton is passed effectively with no loss of momentum. In the Climate Club, the team must work together to benefit from each other's best practices, develop new technologies and possibly implement measures such as carbon adjustments at borders to prevent carbon leakage. This could pave the way to a harmonized and consensual approach to global industry decarbonization. The Climate Club, if it performs well, could potentially transform climate diplomacy by creating a forum for countries to agree on their climate targets, share the burden of innovation, and support each one's efforts in achieving the transition to a low-carbon economy. The Climate Club has the potential, also, to increase global climate ambitions and set precedents that will incentivize collective actions which go further than traditional environmental agreements.

The Climate Club represents progress in the fight against global warming and keeping the world temperature increase below 1.5°C. To become a functioning and effective entity will require fulfilment of several conditions related to the level of participation in the specific actions that will be proposed to meet climate objectives. The economist Nicholas Stern's independent report to the G7 (Stern *et al.* 2022) set out the Climate Club's founding principles as:

1. *Inclusiveness*: From its beginnings, the club must ensure participation of emerging countries to ensure that it becomes a collective that truly reflects a diverse array of climate goals and needs, and not a mere extension of G7 policies.
2. *Unified ambition*: Membership must be predicated on solid commitment to achieving the targets set out in the Paris Agreement and the Glasgow Climate Pact. This shared ambition must be aligned to countries' varying sizes and influence to unite systemic powers and smaller nations in a common cause.
3. *Diverse membership*: The club must be welcoming to nations with different priorities and development levels and must accommodate different policy approaches and contexts within its framework.
4. *Support existing frameworks*: The club must support and extend the ongoing Paris Agreement process and adhere strictly to international legislation, including trade regulations, to ensure that climate action strengthens rather than undermines global cooperation and legal norms.

The Climate Club could provide a platform that would accommodate, recognize and integrate countries' various strategies. For instance, Norway is investing heavily in renewable energy and Brazil is focused on conserving the Amazon rainforest; both strategies are critical to global climate change and efforts but require different approaches and support mechanisms.

Furthermore, adherence to trade law is essential, to prevent imposition of punitive tariffs and sanctions that would ignite economic conflicts. A balanced approach would allow for environmental accountability without disrupting international commerce. The compatibility of carbon border adjustments with the ethos of the climate clubs is a nuanced issue, exemplified by the EU's carbon border adjustment mechanism (CBAM).[26] The CBAM, which was implemented to mitigate the risk of carbon leakage by levelling the playing field for energy-intensive and trade-exposed industries, imposes a carbon cost on imports to the EU, aligned to the EU ETS. It aims to ensure that the price of the related carbon emissions is factored into the cost of goods entering the EU market and was imposed to contribute to the transition towards a decarbonized European industry. During its transitional phase, which began on 1 October 2023, CBAM applies to imports of carbon-intensive goods at risk of carbon leakage, such as cement and steel. Importers must report GHG emissions embedded in their imports but are not yet liable for financial payments. The system will be fully enforceable on 1 January 2026 and will require importers to declare and surrender CBAM certificates corresponding to the embedded GHG emissions in imported goods, with certificate prices tied to EU ETS allowance prices.

However, while the CBAM may have effects within EU borders, the incentive it provides to non-EU countries to join the Climate Club, is not clear. On the one hand, joining the club would entail a commitment to net zero emissions in line with the Paris Agreement.

This could lead to a waiver from the EU (or the Club) CBAM tariff, although the implementation of the CBAM is independent of the Climate Club's framework, and there is currently no direct exemption provided to Climate Club members under the CBAM rules. On the other hand, non-members would be liable for carbon taxes only on exports to the Club's member countries. This raises possibly a strategic dilemma for emerging and developing countries between the benefits of membership, if it leads to avoiding the CBAM tariff when exporting to countries in the Club, and the costs related to committing to bring their economies to net zero emissions.

For instance, a country such as South Africa, whose energy sector depends on coal, might find the CBAM tariff a significant deterrent to trade with the EU. Joining the Climate Club could enable both avoidance of these tariffs and support for the transition to greener energy sources. However, countries sceptical about their ability of reaching net zero might be reluctant to join. The success of the CBAM and the Climate Club will depend on their ability not only to regulate but also to incentivize change. What is important is the balance between enforcing the carbon pricing mechanism and provision of tangible rewards from decarbonization efforts. The ideal system must both protect the environment and support the global industry transformation.

In addition, while the EU CBAM is a leading climate policy innovation, its success will hinge on international countries' perception of its fairness and legality. Transparency is paramount. If non-EU countries are going to view the CBAM as a form of concealed protectionism and a mechanism designed to restrict trade, this could trigger retaliatory actions, such as import restrictions, from these countries. The WTO General Agreement on Tariffs and Trade (GATT) Article XX includes provisions that allow WTO members to deviate from standard trade rules in the interests of environmental conservation. However, these measures are limited by strict criteria: deviations must be rationally structured, backed by solid evidence and be non-discriminatory. Thus, clear justifications and a high level of transparency are required to support any deviation from the rules. However, theoretically, Brazil (a major exporter of agricultural goods) might consider the EU CBAM to be opaque and unjustifiably discriminatory and targeted at Brazil's exports. This could lead to trade disputes and the erection of new trade barriers.

To avoid potential conflicts, the CBAM must be designed and administered such that its environmental goals are unequivocal, and it is clearly not designed to protect EU industries. This will require clear communication of how the mechanism works, transparent sharing of the environmental data driving its policies and commitment to equitable treatment for all trading partners. The CBAM must be seen to be both a climate action tool and a measure that is consistent with the collaborative spirit of international trade law.

The EU has tried to ensure that the CBAM is in line with WTO regulation,[27] but in certain areas, particularly the most favoured-nation (MFN) rule[28] which mandates equal treatment among all WTO member countries, scrutiny is continuing. There is concern that differential treatment of imports based on their carbon content could be construed as discriminatory and in conflict with the fundamental WTO tenet that prohibits the favouring of one WTO member over another. Legal experts emphasize that CBAM adherence to WTO rules will depend heavily on how it is applied in practice and, especially, in terms of differentiation between domestic and imported products (Sapir 2021). The non-discrimination clause, a core aspect of WTO agreements, requires that any domestic policy applied to international trade must treat foreign suppliers on a par with domestic ones, without undue bias.

China is among other powerful states in expressing scepticism about the CBAM and suggesting that it is a veiled form of protectionism. These countries have called on the EU to ensure strict compliance of the CBAM with international trade law, which highlights the importance of ensuring fair competition and avoiding unilateral measures that might disrupt global markets. As the CBAM moves towards full implementation, the EU will be faced with the complexities of these legal and diplomatic challenges. It will be required to demonstrate that the mechanism is a genuine climate policy instrument and not a tool allowing economic protectionism. The ongoing discourse and debate are evidence of the intricacies involved in balancing pursuit of environmental objectives and respecting established global trade rules.

An alternative cooperative strategy to treat global CO_2 emissions would be the implementation of a global minimum carbon price, as suggested by William Nordhaus and also supported by the International Monetary Fund (Nordhaus 2013). Although the current average global carbon price is modest, at around a few dollars per tonne, the High-Level Commission on Carbon Pricing in 2017 proposed a carbon price of $50–100 by 2030 as being more effective and better aligned to the objectives of the Paris Agreement. The proposal for a tiered carbon pricing system would allow account to be taken of the varying levels of economic development among nations and would adhere to the principle of "common but differentiated responsibilities". Under this system, countries would be classified into tiers, based on their development status, with each tier assigned a specific carbon price threshold. For instance, the advanced economies would be in the top (highest carbon price) tier to reflect their greater financial capacity and historical emissions. Emerging economies would be positioned in a mid-tier, reflecting the balance between their developmental status and environmental responsibilities. Least developed countries would be in the lowest tier, to recognize their limited contribution to global emissions and their greater need for support for economic growth.

A tiered pricing structure could reduce the divide between the developed and developing nations by ensuring that all countries contributed to the global effort to reduce emissions but were not burdened by an equal and unfair cost. It would incentivize nations to innovate and invest in cleaner technologies, aimed, ultimately, at decarbonizing the global economy. The implementation of a minimum global carbon price would require international consensus and careful calibration to address economic disparities and ensure fairness. The success of such a strategy would depend on its acceptance by the international community and its ease of enforcement across different jurisdictions, while not impeding the economic prospects of developing nations.

In the contemporary intricate geopolitical landscape, the prospects of achieving global accord, akin to the landmark 2015 Paris Agreement, seem increasingly remote. The 2023 COP28 summit in Dubai, while achieving some critical, albeit incremental, advances, is a stark reminder of the barriers to substantial progress. The conference's achievements underscore the difficulty inherent in securing unanimous support for a multitude of issues from a diverse assembly of national actors. The consensus-based COP approach allows for a small group of countries – or a single nation – to veto a collective agreement. While democratic and inclusive, this often means that the ambitions of the many are curtailed by the objections of a few. The requirement for consensus can lead to a lowest common denominator outcome, in which final agreement reflects not the collective ambition, but rather a threshold acceptable to the least willing parties.

Imagine an orchestra trying to play in harmony without a conductor. If just one section of the orchestra is out of sync, the entire performance is ruined. Similarly, in the COP negotiations, the discordant note of just one country can prevent the adoption of measures that require unanimous agreement. This dynamic was evident in the negotiations in COP28. The challenge lies, also, in not just achieving a consensus, but translating the agreed principles into actions palatable to all. While COP remains a critical climate change forum, the achievement of significant and sweeping agreements may require innovative approaches to circumvent or mitigate the veto power of individual countries.

Taking hope from successful agreements

Should we be optimistic about the potential of international collaborative efforts to address environmental crises? It must be stressed that there are historical examples of agreements that resulted in remarkable success, which could possibly serve as both inspiration and a template for tackling the current climate challenges. A good example here is the global treaty that addressed the alarming depletion of the ozone layer, especially above Antarctica, which acts as the Earth's sunscreen and absorbs most of the sun's harmful ultraviolet radiation. The thinning of this layer – due largely to the widespread use of chlorofluorocarbons (CFCs) in everyday products, such as aerosols and refrigerators – increases skin cancers and cataracts, for example, and causes potential damage to marine ecosystems, with the capacity to disrupt the entire oceanic food web, and harm agricultural productivity through its effects on plant growth. The international response to this challenge, which culminated in the phasing out of CFCs, has been effective. According to a 2022 UN report, these efforts have put the ozone layer on course for recovery (WMO 2022).

The achievement is attributable largely to the Montreal Protocol, which was adopted in 1989 and was a milestone in environmental protection. It has resulted in the phasing out of nearly 99 per cent of ozone-depleting substances such as CFCs. Projections suggest that the continuation of current policies could result in recovery of the global average of the ozone layer to its 1980 levels by 2040. It is also expected that the thinner Arctic ozone should recover by 2045, and the Antarctic ozone layer, where the depletion was first observed, by around 2066.

The success of the Montreal Protocol was due to ground-breaking research. In 1974, Mario Molina and F. Sherwood Rowland from the University of California, Irvine, published a seminal paper in *Nature*, highlighting the risks that CFCs posed to the ozone layer. They described the destructive chain reaction triggered by the chlorine atoms in CFCs that was happening in the stratosphere and jeopardizing the protective ozone layer. Despite resistance from chemical companies and other firms whose products contained CFCs, Molina and Rowland's work and advocacy led, eventually, to a paradigm shift in public consciousness and policy.[29]

Agreement about the importance of and threats to the ozone layer had emerged following a National Academy of Science review published in 1976. In 1985, a scientific team, led by Joseph C. Farman at the British Antarctic Survey, identified significant damage to the ozone over Antarctica. This resulted in global action and the signing of the 1987 Montreal Protocol by 56 countries to phase out CFCs.

The Montreal Protocol stands as a paragon of international agreement, with the distinction of universal ratification by all UN-recognized nations. Through the Protocol, almost all ozone-depleting substances have been phased out. Additionally, the Protocol has indirectly aided in combating climate change by limiting substances that are also powerful greenhouse gases. Despite its success, challenges persist, such as the fluctuating size of the ozone hole and concerns over compliance with the ban on new CFC production. Moreover, the transition away from CFCs has introduced alternative chemicals that inadvertently contribute to global warming. Nonetheless, the overall impact of the Montreal Protocol marks it as a seminal environmental achievement.

Reducing GHGs is a much greater challenge than reducing the use of CFCs. CFCs were involved in a very small proportion of overall economic activity – about 0.1 per cent of US GDP in the 1970s. Thus, the shift away from use of CFCs was less difficult. However, GHGs are emitted by almost all modern industrial and agricultural productions. Also, while CFCs are a distinct group of marketable products, GHGs are not a product that is sold on the market, which makes assessment of their economic relevance more difficult.

For example, if we ascribe a carbon price of $100/ tonne to the 57.4 billion tonnes of CO_2e GHGs emitted at the global level annually, as a form of carbon pricing, the cost of these emissions would be around $5.7 trillion

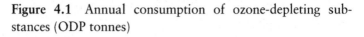

Figure 4.1 Annual consumption of ozone-depleting substances (ODP tonnes)

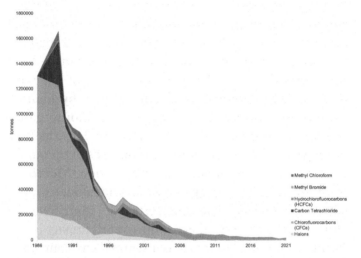

Note: ODP tonnes account for the quantity of gas emitted and how "strong" it is in terms of depleting ozone. The negative consumption values for gasses in some years occurs if gases produced in previous years (stockpiled) have been destroyed or exported.

Source: Our World in Data.

or approximately 6 per cent of global GDP. In addition, this much greater cost compared to CFCs, accounts only for the environmental impact of GHGs, not the value of the goods and services whose production has led to these emissions. This highlights the embedded nature of GHGs in the global economy and their relevance in economic terms.

The implementation of the CFC regulation resulted in a series of innovations. In the US, industry giants, such as DuPont, which initially had resisted phasing out the use of CFCs in aerosols and other areas, developed alternatives (e.g., hydrochlorofluorocarbons or HCFCs), which were promoted during Montreal Protocol discussions.[30] This shift was enabled by the fact that the alternative was in the same industry domain and was patentable, which provided an economic incentive for its production. The transition away from GHG-emitting processes is more complex. While significant progress has been achieved in renewable energy technologies, such as solar, wind, hydroelectric and geothermal power, their adoption is not at a level that will ensure global net zero emissions. Also, there are some industrial processes, such as steel and cement production, where low-emissions alternatives are not yet viable, which emphasizes the need for continuous technological innovation.

For example, whereas in the 1980s, it was relatively easy for refrigerator manufacturers to switch from CFCs to using a less harmful coolant, for some modern industries the task is complicated. The energy sector, currently heavily reliant on fossil fuels, will have to move to use renewables, which will be a monumental shift and affects everything from power plants to cars. Similarly, agriculture, which produces GHG emissions through livestock activities and fertilizer use, will need to revolutionize its practices to reduce the agricultural carbon footprint. Thus, reducing GHG emissions does not involve just replacement of one product or process with another, it involves a transition that will affect the foundations of the global economy.

Critical to the success of the Montreal Protocol was its financial and technical framework to support developing countries. For example, the Multilateral Fund to Support Implementation of the Montreal Protocol, by providing financial and technical support to developing countries, enabled adherence by more countries. It defrayed the costs involved in transitioning away from ozone-depleting substances (ODS) for poorer countries. It supports equitable progress by providing assistance to those countries identified under Article 5 of the Protocol as lower producers and consumers of ODS.[31] To date, 147 out of the 196 parties to the Montreal Protocol qualify for this support. The Multilateral Fund operates on the basis of "common but differentiated responsibility", which underlines that environmental stewardship is a collective duty, with varied expectations based on national developmental status and capabilities.

The Multilateral Fund is inclusive, and decision-making is shared equally between the industrialized nations and those identified by Article 5. Its executive committee reports annually on the Fund's operations to the Meeting of the Parties to the Montreal Protocol. Practical application of Multilateral Fund projects in the relevant countries is facilitated by the United Nations Environment Programme (UNEP), the United Nations Development Programme (UNDP), the United Nations Industrial Development Organization (UNIDO) and the World Bank. These agencies operate under agreements with the committee to ensure effective use of the Fund's resources. Additionally, up to one-fifth of the contributions from donor countries can be allocated through their respective bilateral agencies for projects that fall within the Fund's scope. This provision allows donor countries to directly engage in and monitor the environmental projects they finance. Donor countries contribute to the Fund every three years, which ensures continuous support for developing nations' phasing out of harmful substances to reduce damage to the ozone layer.

Between 1991 and 2005, the contributions to the Fund amounted to $3.1 billion. Funding can be used to finance the conversion of manufacturing processes, personnel training, fees related to the use of patented technologies and establishment of national ozone offices. Funding between 2024 and 2026 is estimated to be $1 billion, the highest level of financing since the Multilateral Fund was established. This funding is critical for progress towards reduction of use of Hydrochlorofluorocarbons (HCFCs) and Hydrofluorocarbons (HFCs) in line with the Montreal Protocol Kigali Amendment and the target of reducing their use by 80 per cent by 2049. However, the allocation of just over $300 million yearly for the next three years, while substantial, is a long way short of the funding required for global climate action to mitigate climate change.

The success of the Montreal Protocol is attributable to its specific focus on protecting the ozone layer to reduce ultraviolet radiation, a health hazard. Reducing CFC use had only a small economic effect and was minimally disruptive to industry. The establishment of the Multilateral Fund involved a large coalition of countries and provides financial and technical support for the transition away from ozone-depleting substances.

Similar initiatives in the area of environmental conservation with a single focus include the 1966 ban on blue whale hunting, which prevented the species' extinction. The International Whaling Commission ban reinforced the 1946 International Convention for the Regulation of Whaling, aimed at protecting all whale species, including the blue whale. The effort related to protecting blue whales is a good example of the power of global agreements. From an estimated 250,000 blue whales in the Southern Hemisphere, industrial whaling reduced this species almost to extinction. In the early 1960s, over 66,000 whales were being killed in a single season. Following the ban, there has been a slow, but positive recovery and between the end of the 1970s and the early 2000s, blue whale numbers were estimated to be around 2,300, with an annual increase of above 8 per cent. Nevertheless, this number is a fraction of historical populations and is the result of over-hunting and the slow breeding rate of these marine giants.

While the task of reducing GHG emissions is enormous, both these successful examples of international agreements perhaps leave room for cautious optimism. Protecting the ozone layer and the blue whale population were major challenges, but history has shown that concerted global efforts can yield significant results.

However, reducing GHG emissions will be a Herculean task; it is hoped that significant reductions can be achieved on an ongoing basis. In this context, China and India, and their major contributions to global emissions, are critical. The long-term sustainability of the planet will be affected by these and other countries' developmental trajectories. The free-rider problem, involving some countries postponing their transition to a greener economy by passing the costs to others, is also crucial to overcome. However, large and influential nations, such as China and India, recognize that their population size and rate of economic growth makes such strategies imprudent. They understand that their active participation is essential to achieve net zero. The objective is not just national economic success; it is also about ensuring a viable future based on successful international climate dialogues.

Cooperation or competition?

Although achieving further global agreements on emissions reductions will be difficult, significant progress is possible. First, both China and India are net energy importers; the benefits of energy independence and lower costs based on renewables are clear for these two nations. Greater use of solar and wind power, for example, would mean reduced variable costs over time compared to fluctuations in the prices of imported fossil fuels. Both energy sources outcompete fossil fuels, making the shift to them both environmentally strategic and economically sound. Second – and linked to the first consideration – renewables offer economic advantages. As these technologies become more affordable and efficient, their adoption is likely to accelerate, reducing both the need to rely on external energy sources and the financial burden associated with importing energy.

This suggests that the transition to renewables will be in line with the economic interests of rapidly developing countries.

Finally, the race to achieve technological leadership in climate solutions is critical for future economic and geopolitical advantage. Achieving net zero, although it will take time, is vital for human survival, and the green technology leaders will be able to capture new markets and achieve economic influence. Currently, China dominates the electric batteries and vehicles, and solar panel production markets. In the US, the 2022 Inflation Reduction Act has worked to support electric vehicles companies, such as Tesla, and to provide incentives for green technology developments, including green hydrogen. These initiatives will be essential for the global shift to use of sustainable energy and is likely to reshape economic leadership.

For instance, the race to develop green hydrogen technology is likely to become significant in the fight for economic supremacy in the coming years. A cost-effective method for producing green hydrogen can provide a considerable strategic advantage and, potentially, could revolutionize the energy sector by offering a sustainable, zero-emissions alternative fuel. This competitive dynamic might catalyse innovation as effectively as international environmental agreements, if not more so.

For instance, several clean energy technologies, including wind turbines, electric vehicles, and battery storage systems, rely on copper, lithium, and rare earth elements, the latter mined and processed mainly in

China. This is raising increasing concern in Europe and the US. As a result, Western governments – somewhat belatedly – have begun investing in a search for these materials. This would diversify their supply and reduce dependence on a single country, while potentially ensuring more stable and less costly sources of these materials. These and similar strategic investments would also increase national security and promote growth of clean energy industries. An expansion to mining initiatives worldwide would also decrease the costs associated with decarbonization and benefit both the relevant countries and the global community. Thus, geopolitical competition in green technology could spur progress towards a more sustainable and economically resilient global energy landscape.

Based on current levels of global emissions, further temperature increases seem inevitable and will be accompanied by more frequent and more severe climate events. This underlines the need to invest in adaptation strategies which, unlike emissions reductions, are localized and benefit the investing party directly. They also avoid the free-rider problem that complicates global emissions agreements. Adaptation measures, such as relocating buildings away from vulnerable coastal areas, yield immediate, tangible benefits for the relevant communities by protecting against the damage and economic losses associated with sea-level rises, for instance. This direct causality between investment and benefit makes adaptation a more straightforward decision for individual actors and communities since the consequences of inaction will be borne locally. Adaptation efforts do not rely on international consensus and do not result in spillover effects.

However, the ultimate aim must be net zero emissions. Adaptations respond to climate impacts, but do not address the root cause. All countries of the world and, especially, those responsible for historical emissions and with the most resources, must work on reducing emissions. The next chapter discusses some practical strategies and actions related to achieving public support for a sustainable transition.

5

Preventing the greenlash

"It always seems impossible until it's done".
Quote attributed to Nelson Mandela.

At the beginning of 2024, farmers across Europe began protesting against new policies and proposals to reduce the tax allowances they receive for diesel, the levels of nitrogen oxide and methane they are allowed to produce, and against the rising costs of fertilizer and other bureaucratic barriers to their businesses. Despite agriculture accounting for approximately 10 per cent of global greenhouse gas (GHG) emissions, on the face of it, the individual European farmer's contribution is limited. While the farmers' protests have brought significant attention to the economic pressures faced by the agricultural sector, they have also highlighted the complex challenge of balancing environmental policy goals with the economic sustainability of farming in the EU. The European Commission and national governments have been tasked with addressing these challenges in a manner that ensures the just transition towards more environmentally friendly agricultural practices without compromising the financial stability of those who work the land. However, the mishandling of farming policies has led to protests and blockades, supported by political groups such as the Dutch Farmers' Party, and a risk of escalation to a full-scale green backlash.

In other sectors, European climate policies have been integrated progressively, as is the case for the electricity industry under the EU Emissions Trading System (EU-ETS). The EU-ETS has effectively reduced emissions by setting a cap on the total amount of greenhouse gases that can be emitted by installations covered by the system. The cap is reduced over time, so that total emissions fall. The gradual reduction has incentivized the electricity sector to invest in sustainable practices and technologies, in line with a smooth transition towards a low-carbon economy. This gradual approach aligns with environmental goals without sacrificing sectoral competitiveness. By providing a clear and predictable regulatory framework, the EU-ETS has enabled power companies to plan and invest accordingly. Investments in renewable energy sources and improvements in energy efficiency across Europe have been supported by these clear, long-term signals.

How environmental policies are enacted plays a critical role in their success and their public support. Overall, across Europe, there is a perceived imbalance in how climate policies are applied to different societal groups, with a need for equitable measures that ensure all contributors to GHG emissions, regardless of their income or industry, are similarly incentivized to reduce their carbon footprint. In this regard, levying carbon taxes for example on high-emission vehicles, energy-intensive properties, private jets, and yachts, therefore ensuring that those with greater means and higher emissions contribute proportionately to climate solutions, would provide a strong signal that policymakers are trying to allocate the cost of the transition fairly and according to means. From the perspective of European farmers in France, Germany, Italy, the Netherlands, and likely elsewhere, resistance to tax increases justified by environmental concerns seems sensible. They perceive themselves to be the first in line to bear the full brunt of such policies when others seem to have more leeway. Whether this is true or not, although important, misses the point that environmental policies must be perceived to be fair in order to be accepted.

A farmer already struggling with thin profit margins might well find that the introduction of new environmental taxes and requirements would risk their entire livelihood. In contrast, other sectors with more significant emissions may not experience the same urgency, as their regulations are still in the pipeline or have longer timelines for compliance. This discrepancy can seem like an inequitable push, asking those with fewer resources to act more quickly than those with more substantial means and broader impact. The point is that how policies are introduced, explained, and set in the framework of other policies, needs to be handled and communicated better. There also needs to be a plan and time for adjustment. The problem is that action does need to be taken at some point soon if we are to have a hope of changing the current trajectory of emissions.

The ubiquity of emissions in daily life

The pervasiveness of GHG emissions in our daily existence cannot be overstated. To address this challenge effectively, it is essential to recognize that our carbon footprint extends far beyond the obvious sources; it permeates every facet of modern living. Most products we consume and the activities we partake in are intrinsically linked to fossil fuel consumption. The path to a sustainable future requires a transformative shift in our consumption patterns, transportation methods and overall lifestyle choices towards low or zero-emission alternatives (see Berners-Lee 2010).

Achieving a net zero global emission status mandates a reduction of almost 60 billion tonnes of CO_2e emissions annually, which is a monumental task. Because of the overwhelming scale of the climate crisis, individual and familial actions hold value but won't get us very far in the overall scheme of things. The primary sources of household emissions include car and plane travel, meat and dairy consumption, and home heating. These activities are deeply woven into the fabric of our lives, making drastic reductions by individuals difficult. Even a simple daily routine – like commuting to work and buying groceries – contributes to the emission tally, and the cumulative effect of these seemingly minor actions is staggering when scaled to a global population.

The problem is that even if households were to eliminate all their CO_2 emissions – a formidable task on its own – the larger issue would persist. For instance, consider the United States, where the Environmental Protection Agency (EPA) has identified that residential energy use constitutes about 20 per cent of the nation's CO_2 emissions. While this is significant, it is important to note that even if American households could miraculously reduce their emissions to zero, this would equate to only a 3 per cent reduction in global emissions. Such an example starkly illustrates that while individual efforts are part of the solution, they cannot shoulder the burden alone. Systemic change, involving all sectors of the economy, is indispensable for meaningful progress in mitigating global greenhouse gas emissions.

As we have already discussed, to articulate a clear and effective climate strategy, it is necessary to embark on a comprehensive electrification of our energy demands. This includes the domains of transportation, and the heating and cooling of buildings, as well as the production of goods and services. The cornerstone of this electrification must be the generation of electricity predominantly from renewable sources, such as solar, wind, hydroelectric and, most probably, also nuclear power.

For sectors where electrification poses greater challenges, such as heavy transport (including trucks and shipping) and aviation, alternative fuels come into play. These alternatives include sustainable biofuels and green hydrogen, which is hydrogen produced using renewable energy rather than fossil fuels. In addition to these measures, it may be also crucial to implement negative emissions technologies. These technologies are essential to counterbalance the greenhouse gases emitted by industries that are currently reliant on carbon-intensive processes, such as the production of cement, steel and fertilizers, where emissions-free alternatives are not yet fully operational at an industrial scale. Finally, the scale of implementation is imperative. Climate change is a global issue that requires international action.

In this context, individual actions aimed at environmental and climate improvement often produce a "feel good" effect more than they contribute to the cause. Opting for plant-based or synthetic alternatives to a traditional beef burger, choosing public transport or bicycles over cars, and supporting local produce to reduce transport emissions are all commendable choices. Significant investments, like installing solar panels or replacing traditional heating systems with electric heat pumps, which have the potential to run on renewable energy, offer more substantial impacts. Such steps are beneficial and when they are part of larger movements, they can have a tangible impact. However, it's important to recognize that their impact can vary, and they come with financial costs that may not be feasible for everyone. Less affluent individuals and those in poorer regions often face barriers to adopting these measures. Hence, while encouraging these actions, we must also strive to address the economic and infrastructural disparities that limit their adoption, ensuring that sustainable options are accessible and affordable for all.

Modern life is intricately intertwined with activities that contribute to CO_2 emissions. Our transportation systems, the materials we use daily such as plastic, cement, and steel, the production of agricultural fertilizers, and even the textiles in our clothes, such as polyester derived from petroleum, are all part of a complex web contributing to the annual emissions. The stark reality is that, within the current infrastructure, living a "normal" life inevitably means participating in a system reliant on fossil fuels. Despite individual changes towards a more sustainable lifestyle, such efforts, although well-meaning, are a drop in the ocean when measured against global emissions.

Confronting this reality that individual environmental actions alone cannot surmount the colossal challenge of achieving net zero, can lead to "climate anxiety" – a pervasive sense of helplessness as we stand on the precipice of potentially catastrophic changes to our planet. Young people are increasingly afflicted by the fear that global warming is advancing inexorably, transforming earth's climate in ways that amplify the frequency and severity of natural disasters.

That anxiety is compounded by a growing body of scientific evidence affirming that this warming is largely driven by human activities (see Chapter 1). The troubling concurrence of extreme weather events with scientific forecasts only adds to the certainty that without swift and decisive action, we may be courting disaster. Yet, amidst this urgency, there remains a palpable absence of consensus on a definitive course of action, further fuelling the collective unease about our environmental future.

Admittedly, altering our energy consumption habits and reducing emissions is a formidable task and even if there is the will, there is a dearth of viable low-emission alternatives. This situation is compounded by the varying levels of environmental commitment across different societies. Meanwhile, it is crucial to acknowledge the aspirations of hundreds of millions of people in developing economies like China and India, who dream of a better standard of living, of owning cars and living in comfortable homes. This global disparity in welfare levels and environmental impact underscores the need for a collective, inclusive approach to climate change – one that bridges the gap between developed and developing nations and fosters a transition to sustainable practices that are fair and accessible to all.

This book has highlighted the significant strides made in the past decades towards a sustainable future, bolstering our confidence that a climate transition is within reach. Yet, it is crucial to understand that the onus does not lie solely on individuals: governments must amplify our efforts with comprehensive regulatory measures that enable and expedite these changes on a much larger scale.

Businesses, much like individuals, encounter numerous challenges in their quest for decarbonization. More resolute political action is needed to enact transformative policies. As it stands, we find ourselves at a critical juncture, akin to a group of climbers stuck at base camp at the foot of a mountain. We have made the necessary preparations, but we must embark on our ascent before time escapes us, and the window to achieve our climate objectives closes with the onset of "winter" – the point of no return. Yet, to embark on this critical journey, we need a collective conviction, a guiding force to unite us in the belief that by sharing the burden equitably, we can surmount the impending challenges for everyone to reach the summit safely.

Public support for climate policies

There has been lots of research into fostering public support for climate policies. A broad key finding has been that the public's perception of how fair and effective these policies are is of crucial importance. Therefore, for a climate policy to gain widespread support, it is essential that it is not only seen as justly distributed across different sectors of society, but also viewed as a genuine driver of environmental progress (Berquist *et al.* 2022).

In the case of carbon pricing, demonstrating the effectiveness of carbon taxes through clear environmental outcomes – such as quantifiable reductions in emissions – should elicit a positive response from the public. Similarly, transparency in the reinvestment of revenues into community renewable projects or in subsidizing energy bills for low-income households would undoubtedly strengthen public endorsement.

By addressing these perceptual dimensions, policymakers can significantly improve the public buy-in necessary for climate policies to succeed. Further studies have placed a spotlight on the power of strategic communication in influencing public opinion. When the benefits and efficacy of a policy are clearly communicated, showcasing how the revenue will be used to promote social equity and environmental progress, it can greatly enhance public support. Therefore, policymakers need to focus not just on the design of carbon pricing mechanisms but also on effectively conveying their benefits and progressive nature to the public, ensuring a deeper understanding and stronger alignment with the policy's objectives (Dabla-Norris *et al.* 2023).

The issue of fairness in the distributional effects of green policies is central to their political acceptance and cannot be stressed enough. When policies are thought to unfairly load burdens onto certain groups while favouring others, public support tends to wane. It is therefore vital that green initiatives are designed and communicated in a way that ensures an even-handed distribution of both their costs and benefits. To achieve broad-based support, policymakers must ensure that these measures do not disproportionately impact disadvantaged communities and that the benefits of a greener economy are shared equitably across society. Only then can such policies garner the necessary consensus to move forward (Colantone *et al.* 2023).

A concrete example of this can be seen in the implementation of renewable energy subsidy programmes that are structured to support both the industry and the consumer. For instance, in Germany, the "*Energiewende*" (energy transition) is a comprehensive policy initiated primarily in response to the 2011 Fukushima incident and the subsequent shift away from nuclear energy.

The policy's cornerstone is the large-scale deployment of renewable energy. Initially, the feed-in tariff (FIT) scheme was applied across various energy sources to ensure a fixed price for electricity produced by renewable sources fed into the grid, encouraging a sustainable energy transition. Over time, the FIT has evolved. Currently, large-scale renewable energy projects, particularly offshore wind farms, are frequently awarded through competitive auctions where developers bid on the compensation they would receive per kilowatt-hour (kWh).

The FIT guarantee extends for two decades and is adjusted based on projected costs and the volume of renewable energy installed in a given year. The financial burden, previously distributed among all electricity consumers at approximately 5–10 cents per kWh, contributing to an annual sum of €25–30 billion, has shifted since 2022. The German government now absorbs this cost, lifting the levy from consumers.

The *Energiewende* is an example of how a green policy can been structured with the aim to achieve wide public acceptance. However, the policy's equity has been a point of contention. While it aims for broader public benefit and stakeholder engagement, two main criticisms arise: the exemption of energy-intensive industries and the regressive nature of the cost distribution, given the inelastic demand for electricity.

Indeed, when the costs of a policy like FIT are spread across all users of electricity, everyone pays the same amount extra per unit of electricity regardless of their income level. Since low-income households spend a larger proportion of their income on essential goods and services like electricity, any flat rate increase in price due to policy costs affects them more. This is why the policy was considered to be regressive: it was imposing a larger relative cost on those with lower incomes, as they could not reduce their consumption of electricity to offset the higher prices, leading to a greater financial strain. These aspects raised questions about the fairness and socio-economic impact of the *Energiewende*, suggesting room for policy refinement to address these concerns (Andor *et al.* 2022).

Gaining the public's support for green policies is feasible. Fairness is a critical factor. When individuals perceive that they are being treated equitably, they are more likely to engage with and support policies that require personal sacrifices. At the heart of this effort is balancing immediate, tangible costs with the pursuit of long-term environmental goals.

Policy priorities

A series of strategic policy goals have been identified as critical on a worldwide scale for the near term. In what follows, the various challenges and impediments that could potentially hinder the successful implementation of these policy measures are explored, alongside a range of feasible solutions and strategies aimed at overcoming these barriers, facilitating the effective realization of our global policy objectives.

Extension of carbon pricing

Revisiting the theme of carbon taxes, it is essential to recognize the potency of this instrument, despite the political challenges associated with its implementation. Carbon taxes serve as a market-corrective mechanism, realigning the cost of emissions with their environmental impact. By levying a price on emissions, they inherently promote a shift in investment towards less carbon-intensive activities, incentivizing innovation and adoption of cleaner technologies without heavy-handed governmental intervention in the private sector's operational decisions.

Such a tax creates a financial impetus for emitters to reduce their carbon footprint, fostering a more sustainable market equilibrium. It is a policy that respects market dynamics while steering them toward a greener future. Moreover, when designed thoughtfully, carbon taxes can be structured to mitigate their impact on consumers and businesses, particularly those most vulnerable to cost increases. By incorporating measures like revenue redistribution or tiered tax rates, the goal is to balance fair economic growth with environmental stewardship, making this policy not just a theoretical ideal but a pragmatic solution for transitioning to a low-carbon economy.

In contrast, when a government elects to provide incentives to promote certain technologies or industries, such as electric vehicles or solar energy, it inherently chooses winners in the market. While these incentives can indeed spur an industry to profitability and consequently draw private capital, governments may not always possess the requisite foresight or market acumen to design the incentives in the best way and to determine the most efficient allocation of private investment. They face the challenging task of predicting future market trends and technological advancements, a process fraught with uncertainties.

By selecting specific sectors for incentives, there is a risk of misallocating resources or overlooking potentially more deserving or innovative sectors. The principle of market efficiency suggests that investors, driven by profit motives and equipped with market knowledge, are typically better positioned to make these investment decisions. Therefore, while government incentives can be beneficial, they must be carefully considered to ensure they do not inadvertently distort the market or hinder the development of other promising sectors. Lots of governments provided incentives to kickstart the green economy, but there are plenty of examples where these policies have not been successful or have been discontinued, for several different reasons.

In some cases, failure is attributable to bad design and communication, such as the UK Green Deal. Introduced in 2013, this scheme was designed to encourage homeowners to make energy-saving improvements to their properties. The idea was that the cost of these improvements would be covered by savings on energy bills over time. However, the initiative faced criticism for its complex nature and the fact that the costs for the improvements were in the form of a loan, the cost of which would be added to energy bills, which many homeowners found unattractive. The uptake was significantly lower than expected, and the programme was ultimately deemed a failure, leading to its discontinuation in 2015. The policy faced several issues, not least a lack of consumer trust and understanding, as well as concerns about the costs being attached to electricity meters, which could potentially make home maintenance more expensive. This example demonstrates how well-intentioned policies can fall short if they are not designed or communicated effectively, leading to a lack of public buy-in and engagement (see, e.g., Howarth & Roberts 2018).

In other cases, policies have had unintended consequences. One notable example was the promotion of biofuels in Europe. The intention was to shift towards renewable energy sources and reduce dependence on fossil fuels. However, the cultivation of biofuels led to unplanned effects, such as changes in land use, which can result in higher overall greenhouse gas emissions than those they were supposed to replace. Forests and peatlands were converted into agricultural land for biofuel production, releasing significant amounts of carbon into the atmosphere. Additionally, the policy did not account for the full environmental costs of biofuel production, such as biodiversity loss and water usage, which led to criticism that the incentives were, in fact, environmentally harmful. The policy's implementation also highlighted the challenges of accurately predicting the environmental impacts of new technologies and the difficulty in creating incentives that lead to genuinely sustainable outcomes without unintended negative consequences.

Sometimes incentive programmes have used vast sums of taxpayers' money with unclear environmental outcomes. Tax incentives aimed at retrofitting buildings to improve energy efficiency that have been provided by several European governments in recent years, run into tens of billions of euros. While comprehensive impact assessments of these subsidies are yet to be conducted, preliminary observations suggest that they have mainly spurred activity in the construction sector and its associated supply chains. However, this investment in the housing sector might not necessarily equate to significant reductions in building GHG emissions, especially when compared to potential investments in green technologies (see Forni *et al.* 2024).

Furthermore, these initiatives may have inadvertently supported industries with high pollution profiles, such as cement and steel manufacture, which are known for their substantial carbon emissions. Consequently, the immediate environmental impact of these policies could be paradoxically negative. Even in the longer term, the net reduction in household emissions involved in the scheme may be modest. Thus, the actual contribution of these tax bonuses to emission reductions is likely to be limited, raising concerns about the strategic allocation of such substantial financial resources.

Considering the broader economic landscape, the implementation of a carbon pricing mechanism stands out as a more neutral and comprehensive approach. By affecting the entire production chain, carbon pricing internalizes the environmental cost of emissions, encouraging more sustainable practices across all sectors. Beyond influencing behaviour, a carbon tax or a cap-and-trade mechanism also has the potential of generating state revenue. This revenue can be strategically utilized to facilitate the transition to a low-carbon economy. It offers the flexibility to provide compensation for those disproportionately affected by the transition or to finance direct measures that support energy efficiency and other green initiatives.

A compelling case can be made for garnering voter support for policies that put a price on emissions. A promising strategy is to give carbon revenues back to households through dividends, tax reductions, or investments in green local projects that enhance neighbourhoods. Such measures should provide immediate, visible benefits which can offset the adverse impacts of the tax, especially on the most vulnerable sections of the population, and therefore reduce opposition. Recycling the carbon revenues back should ease the burden of transition, making the long-term objectives of sustainability more palatable. This approach leverages the principles of political economy by offering an immediate incentive, softening the blow of short-term sacrifices, and thereby fostering a supportive climate for the necessary transitions ahead.

Carbon revenue recycling, while a seemingly attractive strategy to gain public support for carbon pricing, however, is not a one-size-fits-all solution. The success of such policies can vary significantly based on regional differences, cultural perspectives and political leanings. The literature is divided. Some studies find limited evidence that individual, or household, rebates would increase public support for carbon taxes. For example, they suggest that after the "Yellow Vests" movement, French people would have largely rejected a tax and dividend policy, i.e., a carbon tax whose revenues are redistributed uniformly to each adult (Mildenberger *et al.* 2022; Fabre & Duenne 2022). Others show that, while public opinion is sensitive to the cost attributes of carbon taxes, in some cases opposition to them could be offset by a reduction in income taxes. However, these effects tend to be modest in size, limited to ideological groups leaning more towards the environmental cause, and varied across countries. But there is also evidence in favour of recycling carbon revenues. For example, it could increase support for carbon taxes, if the tax is not too high, and – evidence suggests – under the hypothesis that other countries join forces and adopt similar carbon taxes. In the US, where carbon taxes are not popular, there are also political connotations. Conservatives would be relatively more supportive of a carbon tax if revenues would be allocated toward a tax rebate or deficit reduction (Beiser-McGrath & Bernauer 2019; Nowlin, Gupta & Ripberger 2020; Jagers *et al.* 2021).

However, while carbon tax rebates might gain traction in the short run, do they compromise long-term objectives? From a purely efficiency-driven standpoint, using carbon revenues for more direct emission reduction methods seems more sensible. Investing in R&D, fostering green industries, or even directly intervening in emission-heavy sectors holds the potential for a more rapid and significant reduction in emissions. Such strategies accelerate the shift toward a more sustainable economy. By redistributing carbon revenues back to the people, it might inadvertently be putting the brakes on the green transition. With more money in their pockets, consumers will naturally increase their consumption. In an economy still dominated by high-emitting technologies and not yet fully transitioned to green alternatives, this would mean a higher demand for emission-intensive products and services.

In essence, the dilemma is clear. On the one hand, there is the need to earn public support by ensuring perceived fairness, and on the other, the imperative to drive an efficient and rapid green transition. Striking the right balance requires nuanced policymaking, where immediate public benefits are coupled with a concerted push for sustainable transformation. Policy architects need to consider the economic implications and the socio-political fabric that influences public opinion. We can only navigate the intricate challenges of transitioning to a green economy with this comprehensive approach.

Also, focusing solely on carbon revenue recycling can be a narrow approach when considering the broader picture. Carbon pricing is one instrument in the vast toolkit of mitigation strategies, and not all these tools come with fiscal revenues that can be channelled back to the public. For instance, regulations, standards, incentives, direct public spending for mitigation and adaptation might not generate direct fiscal revenues but can still be pivotal in driving a country's green transition. Moreover, redesigning crucial markets, like the electricity sector, is another way to allocate the benefits of the transition to specific groups. Furthermore, in an age of information overload, the narratives championed by interest groups, lobbyists and incumbents can significantly influence public opinion. The challenge, then, is designing effective policies and ensuring they are communicated authentically and transparently to the public.

In conclusion, relying solely on carbon revenue redistribution as the primary strategy to gain public support might be overly optimistic. It should be considered as a part of a comprehensive suite of policy tools, each tailored to the unique circumstances and requirements of individual nations and their citizens, including both social programmes and redistributive elements.

Expansion of renewable capacity

A notable example of the difficulties of investing in renewable energy in advanced economies is provided by Italy. Despite its significant potential for harnessing solar and wind energy, Italy's growth in renewable electricity capacity has been perplexingly slow. The National Integrated Energy and Climate Plan (PNIEC), updated in December 2023, outlines Italy's ambitious goals: to elevate its solar and wind electricity production from 41,600 gigawatt-hours (GWh) in 2017 to 163,200 GWh by 2030. This target represents a fourfold increase within a span of 13 years.

Yet, the reality on the ground paints a different picture. The pace of renewable energy investment over recent years has been modest, casting doubt on the feasibility of the 2030 targets. Moreover, the Italian National Plan for Recovery and Resilience (PNRR), funded by the EU, presents a worrying incongruence with Italy's ambitious renewable energy objectives. The PNRR's forecast anticipates for 2026 a renewable energy output of about a third of the PNIEC targets for 2030. Moreover, the financial commitment to renewable energy within the PNRR is notably modest: about 6 billion for wind and solar energy by 2026. These allocations are insufficient when weighed against the scale of investment required to meet the outlined targets.

The bulk of renewable energy financing is expected to come from private sources; however, the pivotal role of public investment cannot be overstated. It is crucial for government to not only accelerate the deployment of private renewable projects but also to channel funds into upgrading the grid infrastructure. Such enhancements are indispensable for the seamless integration of renewable sources. In the case of Italy, the current trajectory suggests a need for a recalibrated strategy of its public funding commitments.

Italy's current investment strategy shortfall exposes the country to the twin peril of failing to meet its international emission reduction commitments and jeopardizing its energy security. The events of 2022 serve as a stark reminder of these risks. A severe drought significantly reduced the country's hydroelectric power generation, forcing energy providers to rely on expensive gas purchases, as well as oil and coal to a lesser extent, amid an energy crisis precipitated by reduced Russian gas supplies. The inadequate renewable capacity meant that solar and wind energy could not offset the shortfall in hydroelectric output. As a result, Italy saw a decrease in renewable electricity production in 2022 and a huge spike in energy costs. A robust renewable energy infrastructure could have provided a more reliable and cost-stable alternative, leveraging Italy's abundant natural resources like sunshine and wind – free and immune to the volatilities of international politics.

The expansion of renewable energy in Italy is not limited by profitability considerations. Indeed, both photovoltaic and wind power are now cost-effective ventures with significant potential for growth. Italy, for example, has ample opportunities for offshore wind power development. The true impediment to progress is the cumbersome bureaucratic process for greenlighting these projects. This lethargic administrative system often mirrors a broader societal reluctance, encapsulated by the "not in my back yard" (NIMBY) syndrome, where local communities resist the siting of renewable installations nearby.

The challenge of NIMBYism in the context of renewable energy projects is indeed widespread and multifaceted, but solutions are within reach. Current evidence points to the importance of public engagement and negotiation as key strategies in overcoming local opposition. Researchers have underscored the critical role of including local communities in the decision-making process (see, e.g., Richman 2002; Devine-Wright 2009). This is not just about informing the public, but genuinely involving them in dialogue, ensuring their concerns are heard, understood and addressed. It is about building a participatory approach that can foster community buy-in and ownership of renewable energy projects (O'Neil 2020).

Furthermore, the complexity of NIMBY thinking suggests that opposition is not just a knee-jerk reaction to change, but often rooted in deep-seated attachments to local environments and ways of life. Residents' connections to their locale and the impact that renewable energy projects might have on these places must be taken into consideration (O'Neil 2020). This complexity indicates that solutions to NIMBYism need to be as varied and sophisticated as the concerns themselves, tailored to the specific contexts of each community. Additionally, critics challenge us to look beyond the conventional NIMBY label, which may oversimplify the reasons behind opposition to renewable energy installations (Burningham 2000). The NIMBY label doesn't account for the legitimate concerns of residents and can undermine the depth of understanding needed to address these issues effectively. A more nuanced approach is called for, one that goes beyond labelling and seeks to understand the unique circumstances and concerns of each community.

In sum, while NIMBYism presents a considerable barrier to the advancement of renewable energy projects, especially in advanced economies, by reframing engagement with communities as a partnership rather than as a confrontation, developers and policymakers can make significant strides in advancing renewable energy projects in a way that benefits all stakeholders.

Greater investment in energy saving

In the summer of 2022, the European Commission urged member states to reduce their energy consumption by 7–15 per cent. This recommendation was part of a broader strategy to navigate the winter of 2022–23 without resorting to energy rationing, a concern that intensified with the sharp decline in gas supplies from Russia. As the end of 2022 approached and the new year began, a significant reduction in energy consumption had been indeed recorded relative to historical norms. This decrease was partly attributable to the mild winter temperatures and the deterrent of high gas prices, with additional savings achieved through governmental mandates regulating indoor heating levels.

However, these savings were largely the result of circumstantial factors and are not indicative of a permanent shift towards lower energy consumption. Certain energy intensive industries reduced production, but hopefully with an aim to recover it once prices would fall to more sustainable levels. No substantive structural policies were put in place to ensure sustained energy conservation. It is imperative that in addition to temporary measures, we develop and implement long-term strategies to curtail energy use. Such policies rely on enhancing energy efficiency in buildings, industries and consumer products such as appliances. Reducing personal transportation uses in favour of shared options and investing in smart energy management systems are also helpful measures. By doing so, we can secure lasting reductions in energy consumption.

Energy saving constitutes a fundamental pillar in the architecture of emission reduction strategies. The European Commission, within its comprehensive Fit for 55 legislative packages has pinpointed energy conservation as a key objective. The Commission's vision includes a projected 39 per cent cut in primary energy consumption by 2030 relative to 2007 levels, with energy efficiency in buildings playing a crucial role in achieving these figures. This aligns with ambitious targets for renewables, which are expected to constitute 40 per cent of the EU's energy mix by 2030, and a reformed European Emission Trading System (ETS) designed to lower the overall emissions cap and, consequently, increase the private cost of carbon emissions.

Although energy efficiency upgrades are a critical aspect of reducing consumption, their impact is but one piece of a much larger puzzle. It is crucial to recognize that the scope for energy savings extends far beyond the realm of improving efficiency. As we mentioned in Chapter 2, the principles of the circular economy offer vast, untapped potential for conservation. Significant energy reductions can be achieved by extending the lifespan of products, promoting reuse, and rigorously recycling both consumer items and industrial waste. Such strategies not only diminish the demand for energy-intensive production of new goods but also reduce the overall extraction of raw materials. Embracing the circular economy involves rethinking our approach to consumption and waste, ensuring that resources are utilized as efficiently as possible. By integrating these principles into our broader energy savings strategy, we can amplify the impact of our efforts to curb energy use and contribute more effectively to the overarching goal of emission reduction.

The practice of intentionally shortening the lifespan of products is deeply embedded in the capitalist economy's structure, driven by the pursuit of profit. When goods, even those intended to be durable, have shorter life cycles, the manufacturing companies stand to profit. The more frequently a company can introduce new models to supersede the old, the more sales and, consequently, profits it can generate. This is becoming a clear problem, evident to many people. For example, many consumers upgrade their smartphones every few years due to new models with updated features and technologies. This practice leads to other environmental concerns due to the waste of materials and the energy used in manufacturing new devices, as well as the challenge of disposing of the old ones. Recognizing this issue, some governments and organizations have started to advocate for more sustainable practices, such as encouraging the development of modular phones that are easier to repair and upgrade or implementing regulations for better recycling programmes to reduce electronic waste.

But this raises an important question: Why isn't this principle of extending product lifespans applied more broadly to consumer goods? Implementing similar practices across various sectors could significantly reduce waste and energy consumption, challenging the throwaway culture that exacerbates our environmental footprint. It's time to reconsider the principles that govern our production and consumption and to establish frameworks that promote sustainability over short-term gains.

Consider the automobile industry in the United States, where roughly 300 million vehicles are in use. Assuming the average lifespan of a car is ten years, which is close to actual figures, approximately 30 million vehicles are discarded annually. By increasing the average service life of vehicles by 20 per cent, to 12 years, we could potentially reduce the number of cars scrapped each year by 5 million. The energy savings from such a reduction would be considerable, though challenging to quantify precisely given the complexities involved in vehicle manufacturing and disposal.

However, this proposition requires a nuanced approach. In regions with older, more polluting vehicles, prolonging the life of the car fleet could paradoxically result in higher emissions. Therefore, the strategy of extending the service life should be systematically and selectively applied, focusing on those durable goods – such as energy-efficient vehicles and household appliances – that do not compromise environmental standards. This would involve a careful balance between maintaining operational efficiency and reducing the frequency of replacement, thereby conserving the energy and materials used in production.

Car owners are currently free to replace their vehicles as they see fit. Imposing restrictions on how frequently consumers can change their cars may be considered to be an infringement on personal liberties and is hardly practicable. Nevertheless, the introduction of well-designed regulations and incentives aimed at encouraging car owners to retain their vehicles longer is a viable approach. Here are some examples of measures that have been tried to this effect with reference to personal vehicles:

1. *Subsidies for maintenance*: subsidizing the cost of maintaining older vehicles can make it more economically viable for owners to keep them. This could include vouchers for servicing or discounts on replacement parts. In some jurisdictions, programmes have been implemented that subsidize the replacement of key components like catalytic converters or offer free emissions testing, which can help older vehicles remain both operational and compliant with environmental standards.

2. *Certification programmes for used vehicles*: certification programmes can increase the value of used cars, making it more attractive for owners to sell them on the secondhand market rather than scrap them. Certified pre-owned (CPO) schemes often include thorough inspections and extended warranties, giving buyers confidence in the quality and reliability of used vehicles.

3. *Campaigns to raise environmental awareness*: public campaigns can effectively shift consumer behaviour by highlighting the environmental benefits of retaining older vehicles. For example, the "Don't Trash Your Car Unless It's Trash" initiative in the United States aimed to educate car owners about the environmental costs of vehicle production and disposal, thereby encouraging them to consider the full life-cycle impact of their cars.

4. *Trade-in programmes*: some car manufacturers have introduced trade-in schemes that offer incentives for owners to replace their older, less efficient vehicles with newer, more efficient models. While this may seem counterintuitive to the goal of extending vehicle lifespans, such programmes can remove the most polluting vehicles from the road and replace them with cleaner alternatives. There are many examples of this type of policy, like the US "Cash for Clunkers" programme, which aimed to stimulate the economy and encourage the purchase of more fuel-efficient vehicles.

Nudge theory and behavioural economics can also play a significant role in influencing consumer behaviour without limiting freedom of choice. The concept of "nudge" refers to subtle policy shifts that encourage people to make decisions that are in their broad self-interest. It is about making it easier for an individual to make a certain decision but without restricting their ability to choose. One practical example of a nudge policy implemented in the context of extending the use or life cycle of personal cars is the introduction of a car maintenance reminder system by vehicle registration authorities or automotive service providers. In this case, a government agency or automotive service provider implements a digital reminder system that automatically notifies car owners of upcoming maintenance schedules based on the vehicle's age, mileage and model-specific maintenance requirements. This system could leverage existing registration and insurance renewal frameworks to ensure it covers as many vehicle owners as possible. This nudge aims to increase the frequency and timeliness of vehicle maintenance among owners, which in turn helps in extending the life cycle of personal cars. Properly maintained cars are more efficient, safer, and less likely to require costly repairs or replacements prematurely. This not only benefits the car owner but also contributes to environmental sustainability by reducing the need for new cars and the associated manufacturing and disposal impacts.

By implementing such measures, governments can create a regulatory and economic environment that encourages the retention of more efficient vehicles without imposing on personal freedoms. There is a delicate balance between providing incentives for maintaining older vehicles and ensuring that the overall vehicle fleet becomes cleaner and more environmentally friendly over time. Reflecting on these approaches with creativity and a commitment to sustainability could lead to effective strategies for prolonging the life of consumer products. Each policy option would require careful consideration of potential economic impacts, consumer behaviour, and the overall goal of reducing the environmental footprint. Such measures could form an integral part of a broader strategy to encourage sustainable consumption and production patterns, with the end-goal of reducing energy consumption.

Enactment of climate laws

The adoption of a national climate law is a crucial step for countries to concretize their dedication to long-term climate and environmental objectives. A comprehensive climate law would not only codify a nation's climate targets but also establish the guiding principles for all future governmental policies. For example, the Climate Change Act of 2008 in the UK was a pioneering piece of legislation that set legally binding carbon reduction targets, thereby creating a framework for all subsequent climate-related policies. Similarly, the EU's European Climate Law of 2021 aims to enshrine the 2050 climate neutrality target into law, creating an irreversible commitment to this goal.

The public's role in national climate laws is fundamental. Such legislation often includes mechanisms for public accountability, where governments are required to report on their progress, ensuring transparency. For the public, these laws provide a clear direction to the nation's climate efforts, which can guide personal and community action towards sustainability. Climate laws would facilitate the participation of citizens in climate action movements through various means, such as public consultations, contributing to the discourse on policy development through citizen assemblies, and more generally holding policymakers accountable through advocacy and electoral processes. Essentially, climate laws offer a legal framework that both directs, and is influenced by, public engagement and societal behaviour, making individual and collective action a key component of achieving climate objectives.[32]

A climate law should be underpinned by key principles that guide a country's environmental policy and actions. One such principle, which we already discussed, is the

"Do No Significant Harm" tenet, which would obligate governments to abstain from implementing measures that significantly degrade the environment or contravene emission reduction efforts. This principle serves as a safeguard against policies that could undercut the progress towards climate goals. For instance, we have mentioned before that during the energy price surges of 2021–22, European governments enacted widespread measures to alleviate the financial burden on consumers, and that there is a consensus that these interventions were overly broad, reaching beyond the most vulnerable households and businesses. Consequently, the high energy prices, which could have naturally prompted more judicious energy use and investment in renewables, did not fully exert their potential influence. Furthermore, the measures dampened the momentum towards renewable energy by funding the interventions in part through levies on the profits of energy companies, including those producing energy with renewable technologies.

A robust climate law, governed by the principle of "Do No Significant Harm", would have required a more targeted approach, ensuring support reached those in genuine need without diminishing the impetus for energy conservation and the shift to renewable sources. It would encourage the use of market signals, like high energy prices, to drive behavioural changes and investments in clean energy, while also providing a safety net for those who might be disproportionately affected. Incorporating such a principle in a climate law would mean that any future measure that potentially harms the environment, or is inconsistent with emission reduction objectives, must be carefully vetted and avoided. It underscores the necessity for climate policy to be designed not just with economic considerations in mind but also with the foresight to protect and preserve environmental integrity for the long term.

A second principle, which we have already mentioned, is the precautionary principle. This principle is a vital guide in environmental stewardship, particularly when dealing with activities that may alter the environment in ways that are not fully understood. In many cases, the exact consequences of environmental changes on ecosystems or human health are difficult, if not impossible, to determine with precision. For example, quantifying the specific health impacts of varying air pollution levels on respiratory disease mortality, understanding the exact repercussions of certain water pollutants on fish populations, or predicting the full extent of ecological damage from oil drilling in sensitive areas, are complex challenges that may not yield precise forecasts.

The precautionary principle advocates for a conservative approach to environmental management. If there is a plausible risk that an activity could have detrimental environmental or health outcomes, it is prudent to restrict or modify that activity proactively, rather than waiting for absolute scientific proof of harm, which may come too late. This principle operates on the understanding that prevention of harm is more beneficial, both ethically and economically, than remediation after the fact.

In practice, the precautionary principle has informed various international agreements and national laws. It encourages policymakers and industries to consider the potential negative effects of their actions and to take responsible steps to avoid them, even in the face of scientific uncertainty. This might mean investing in more rigorous environmental impact assessments, enforcing stricter controls on potentially hazardous substances, or foregoing certain developments until their safety can be assured. By embedding the precautionary principle in a climate law, governments would underscore their commitment to safeguarding the environment and public health, even when future impacts cannot be predicted with complete certainty.

By embracing the Do No Significant Harm and precautionary principles, we could significantly transform the way industrial products are produced and consumed. One tangible application of these principles could be the mandatory disclosure of the carbon footprint of products, akin to nutritional information on food packaging. Such transparency would enable consumers and regulators to assess the climate impact of different products more accurately. If products were required to clearly display the emissions associated with their production, consumers could make more informed decisions, favouring goods with lower environmental impacts. Moreover, this information could inform policy decisions. For instance, using the precautionary principle, products that result in emissions exceeding a certain threshold could be subject to restrictions or bans, mitigating their potential harm to the climate.

This approach would not only empower consumers but also incentivize manufacturers to reduce the carbon footprint of their products. Companies might invest more in cleaner production technologies or redesign products to be more environmentally friendly, knowing that consumers are increasingly making choices based on sustainability criteria. Furthermore, this policy could lead to the development of a labelling system that rates products based on their environmental impact, like energy efficiency labels on appliances. Such a system could also include incentives, like tax breaks or subsidies, for products that demonstrate superior environmental performance.

A national climate law should also detail clear, actionable targets for emission reductions, renewable energy adoption and conservation efforts. It would also define the governance structures necessary for monitoring progress and enforcing compliance. Additionally, such a law could lay out the mechanisms for public and private sector collaboration, citizen engagement, and the integration of climate objectives into economic and social policy planning. By having a climate law, governments can ensure policy continuity across different political administrations, providing a stable environment for investment in green technologies and giving citizens a clear understanding of the nation's environmental trajectory. It is a way to transcend political cycles and create a legally binding commitment to climate action that can hold future governments accountable to established goals.

In closing, it is essential to circle back to the foundational principles discussed at the outset: fairness, a just transition, and sustaining support for effective climate policies. As we have explored, these elements are not just philosophical ideals but practical necessities for the successful implementation of environmental strategies. Fairness must be the cornerstone of policy frameworks to ensure equitable distribution of both the responsibilities and benefits of climate action, enabling all sectors and communities to participate without disproportionate burden. A just transition is paramount, carefully orchestrating policy shifts to mitigate negative impacts on vulnerable groups. Finally, maintaining public support hinges on transparent communication and demonstrable effectiveness of policies, fostering a shared commitment to the cause. By embedding these principles at the heart of climate policy, we can strengthen societal commitment, drive meaningful action, and sustain momentum towards our collective goal of a sustainable and just environmental future.

Conclusion

The path to net zero is fraught with technical, economic and political challenges.

There is an evident lack of political will currently to enact the robust measures necessary for achieving net zero emissions. It appears increasingly clear that reaching this monumental goal will likely extend beyond the initially targeted year of 2050.

What this underscores is the need for a sustained and intensified effort from the political sphere. Governments must transcend short-term electoral cycles and partisan agendas to prioritize the long-term health of the planet. This calls for a paradigm shift in policymaking, where climate considerations are not peripheral concerns but central tenets of all legislative and executive actions.

The way forward involves a comprehensive mobilization that includes but is not limited to the adoption of stringent climate laws, the integration of principles such as "Do No Significant Harm" and the precautionary principle, and a rigorous re-evaluation of economic systems to encourage sustainable and just practices.

Governments must ensure that the transition is fairly implemented.

Societal engagement is equally crucial; public demand for action can galvanize governments to commit to the necessary transition policies. Political actions are, in large part, a reflection of voter preferences. Unless voters make it unequivocally clear that they will hold politicians accountable for inadequate action on climate change, it may continue to be challenging for leaders to commit to the bold steps required.

Role of governments

The shift towards a sustainable economy is fraught with complex adjustments that have significant political and financial implications across various layers of society. At the core of this dynamic is the reluctance to incur upfront costs, despite the promise of widespread, long-term benefits such as lower emissions and reliance on cleaner, domestic energy. Economic theory illuminates this tension, highlighting how the immediate financial burdens faced by businesses and households – although sometimes mitigated by public funding – yield societal benefits that extend far beyond their individual scope. This is what Mark Carney has defined the "tragedy of the horizon" (Carney 2015).

The tension between immediate costs and future benefits is heightened by another tension, making the road to net zero even more bumpy: the divergence between private costs and private benefits in reducing emissions, which is emblematic of the well-known economic dilemma related to externalities. For instance, while the environmental improvements resulting from one company's green initiatives benefit society, the company may not receive proportional private economic gains. Conversely, if a firm's operations harm the environment, the consequences are often dispersed among the wider public rather than being fully borne by the firm itself. This misalignment can dampen the motivation for proactive environmental stewardship and lead to free riding, a scenario in which individuals or businesses avoid their share of the costs, hoping others will take the necessary actions. Such behaviour is a significant hurdle in the collective journey towards a green transition, as it undermines the shared responsibility essential for achieving our environmental goals. To surmount this obstacle, government policies must align private incentives with public interest, ensuring that the costs and benefits of environmental actions are distributed more equitably. This means not only encouraging positive practices but also holding all parties accountable for their environmental impact, steering the economy toward a sustainable future where the true value of clean air, water and land is recognized and factored into every economic decision.

While the traditional arguments highlighting the disparity between private costs and public benefits in climate action remain relevant, also on the international scale, recent developments compel us to revisit these assumptions. The landscape of climate change mitigation is being reshaped by the swift progress of low-carbon technologies which are increasingly cost-competitive. In regions heavily reliant on fossil fuel energy imports, like Europe, the economic case for transitioning to these technologies is even stronger, not just from an environmental standpoint but also from a cost and security perspective. The transition to clean energy is now as much an industrial imperative as it is an environmental one. Major economies are recognizing the strategic importance of leading in clean technology, with a clear pivot towards implementing industrial policy as they vie for dominance and market share in this burgeoning global green sector. This new paradigm presents a compelling narrative: investing in emissions reduction today is not only a moral and environmental imperative but also an economic and geopolitical priority.

As we navigate the intricacies of the green transition, we must also confront the immutable realities of climate dynamics. The legacy of historical emissions has already inscribed a temperature increase of over 1°C into the annals of our planet's climate history – a change that is largely irreversible. Current emissions are adding to this burden, propelling us towards an even more precipitous climatic future. Even with aggressive cuts in emissions, we are essentially attempting to brake on a downward slope that continues to steepen; our most hopeful scenarios involve stabilizing – not reversing – this temperature rise at new, higher baselines.

This situation presents a stark dichotomy: the investments we make today in reducing emissions are set against the backdrop of a persistently warming planet. It highlights the acute necessity for a dual approach in our climate strategy. Achieving net zero emissions alone is likely to be insufficient. We must also actively reduce atmospheric CO_2 levels through the deployment of negative emission technologies. These interventions must proceed in tandem with ongoing mitigation efforts, rather than as a substitute. The challenge before us is to not only stop adding to the problem but to also engage in the monumental task of reversing the damage already done.

We are at a juncture where clear, credible and consistent public policies are not just beneficial but necessary for the attainment of net zero emissions. These policies serve as the backbone of climate action, guiding and mandating the systemic changes required across industries and economies. While our individual actions – such as reducing waste, supporting green businesses and adopting a sustainable lifestyle – are important, they need to be complemented by, and indeed reliant on, the framework that only effective public policy can provide.

The role of the public

The path to climate action is strewn with obstacles that go far beyond issues of equity and economic externalities. At the heart of these challenges is the variable level of public engagement: despite the far-reaching consequences of climate change, a significant portion of the population remains either unaware or indifferent to it. This apathy is compounded by organizational barriers, such as the daunting procedural maze to greenlight renewable energy initiatives. The "not in my backyard" (NIMBY) syndrome exacerbates this, with local resistance to the siting of renewable projects despite a wider approval for their necessity.

The hurdles are amplified by the influence of powerful interest groups that have a stake in perpetuating fossil fuel reliance, often swaying public opinion and skewing policy agendas to favour traditional energy sources. Added to this mix are contentions over the optimal policy approach to catalyse the transition. The debate oscillates between advocating for carbon pricing as a punitive measure against high emissions and promoting incentives to spur innovation in emerging green sectors.

This intricate web of challenges underscores the imperative for a comprehensive approach in policy-making, integrating market-based tools, regulatory frameworks and support for innovation. However, communicating the intricacies of such sweeping policy changes is a formidable task, requiring a degree of political courage and leadership that is often hampered by these varied pressures. The pressing need is for decisive political action that can cut through these complexities, uniting diverse stakeholders behind a common goal of meaningful climate progress.

How can policymakers forge a connection between the personal costs borne today and the collective rewards of tomorrow? To address this, governments must craft a narrative that resonates with the public, one that articulates the tangible benefits of climate action in terms of health, economic stability and community resilience. Policies must be designed to not only mitigate the immediate burden on individuals but also to highlight the near-term positives that come with sustainable practices. This could involve showcasing the job creation potential in renewable sectors, the health benefits from reduced pollution, and the economic gains from energy independence.

Ultimately, the goal is to weave the thread of sustainability through the fabric of society, ensuring that the larger climate objectives are primary, not pursued at the expense of our diverse needs, but in harmony with them.

It is also critical to acknowledge that we all hold a shared responsibility as environmentally conscious consumers and that we must join efforts in this fight. Each individual decision we make *per se* might not be very relevant, but in the aggregate they can help to send a message to producers and the financiers behind them, building collective pressure that can steer corporate investment toward emission-reducing initiatives. The push for transparency in product labelling to include emission content will empower consumers to make choices that align with their environmental values. This form of advocacy, however, is no substitute for the more fundamental changes that we can achieve through our electoral influence.

To galvanize collective action, we must craft policies through an inclusive lens – drawing communities into the heart of decision-making and transforming climate action from a mandate to a mutual quest for a brighter future. Our era demands policymakers who can merge private interests with public welfare and distil the essence of long-term gains into tangible, immediate rewards. Securing widespread support is critical for enacting the bold climate strategies we need.

Appendix:
the challenges in defining and measuring emissions

Measuring emissions is difficult. At present, only large companies, especially those subject to emission trading systems – such as the European Emission Trading System (EU ETS) – have protocols for measuring them. Most smaller companies do not measure them. Just as we use estimates, not direct measurements, for emissions from transport, housing, agriculture, and so on.

The practice currently divides emissions into three categories:

Scope 1: those produced directly by a production process;

Scope 2: those related to the energy used by the same production process;

Scope 3: all other emissions related to intermediate inputs, their transport, the transport of goods produced by the production process in question and their use.

For example, for a plant producing plastic parts, Scope 1 emissions are those related to the production process of the plastic parts, Scope 2 emissions are those related to the production of the energy purchased and used in the plant, and Scope 3 emissions are those necessary to produce the intermediate inputs to the production process, as well as for transport upstream and downstream of that process. There is a risk of counting Scope 3 emissions twice, once in the plant that produces them directly and once in the plant that uses that product as an intermediate input.

In the taxonomy of emissions, there are also so-called *Scope 4* emissions, i.e. "avoided emissions". Companies may develop new products or processes that produce fewer emissions than their predecessors throughout the production chain but may be reluctant to do so if the new product still leads to emissions. Hence the need to assess whether the new product or process leads to a reduction in emissions, compared to the status quo, throughout the value chain. It is important that "avoided emissions" are recognized and evaluated, so that companies have the right incentives to introduce innovations that in net terms lead to a reduction in emissions.

The possibility of increasing emissions now to diminish them later is also an issue worth considering. Initiatives such as constructing solar and wind energy facilities, bolstering electric grids, or advancing battery production may indeed intensify emissions and investment needs briefly. However, such steps are pivotal for forging a sustainable, low-emission future. A strategic approach would involve quantifying the total emissions offset over an extended period, ensuring that short-term escalations are eclipsed by long-term reductions. To create the right incentives, the net amount of emissions saved over time should be calculated.

There is also a debate as to whether reducing deforestation should be included in "avoided emissions". When trees are cut down, they release the carbon they have stored into the atmosphere. They release more and more quickly if they are burnt. Added to this, the reduction of forests reduces the absorption of CO_2. Overall, therefore, forest felling contributes a significant amount to annual net emissions, estimated at 10–20 per cent. If a country decides to reduce the rate of deforestation, or to reforest certain areas, should it be able to count the avoided emissions towards its emission reduction targets? Of course, it depends on what one considers the baseline to be. If the baseline is progressive deforestation, then stopping it means reducing net emissions and these should be counted as emission reductions. If the baseline is stopping deforestation, then net emissions are not reduced. Obviously, the issue is sensitive, but if countries are to be incentivized to conserve more forests, if not increase them, then the net emission reductions that conservation entails should be recognized.

But there is more. We must be careful how we reduce emissions. Many countries have set ambitious interim targets for 2030. Meeting these targets with short-term solutions could hinder the achievement of net zero on the longer horizon, by 2050. For example, many countries could significantly reduce emissions in the next few years by building gas-fired power plants, which, however, have a long life cycle and could then be an obstacle to switching to renewable energies with much lower emissions, and thus hinder or defer the achievement of net zero. In short, we should avoid ambitious intermediate targets becoming obstacles in the attainment of final ones. If intermediate targets are to be achieved at any cost, there is a risk that they will be more counterproductive than helpful.

These few notes on emissions are intended to make it clear that greenhouse gas emissions are not carved in stone. Even if they are believed to be estimated fairly accurately, many definitional problems remain in calculating their reduction path and progress towards net zero targets. Incentivizing technological advancements in emissions reduction requires that companies are acknowledged for the broader impact their innovations have across the value chain, not just within their own operations. It is essential to recognize the potential of new products and processes to lower emissions in the future, as well as in the present. When it comes to nations, it's equally important to account for emissions that are avoided through strategic decisions, such as expanding forested areas, as a legitimate part of their efforts to achieve emission targets.

Notes

1. The European Commission Special Eurobarometer: Climate Change poses a range of questions on climate aspects to representative citizen samples in various European countries: see https://data.europa.eu/data/datasets/s2954_99_3_sp538_eng?locale=en. In the US, the Pew Research Center conducts surveys that include questions on climate issues: see https://www.pewresearch.org/global/2022/08/31/climate-change-remains-top-global-threat-across-19-country-survey/.

2. "Tipping points" in the climate system are critical thresholds at which small changes can lead to significant and potentially irreversible shifts in the state of the system. These tipping points can result in abrupt consequences such as the rapid loss of ice sheets, significant changes in ocean circulation, or widespread dieback of forests, each of which can have dramatic impacts on global climate patterns and ecosystems. The crossing of these thresholds is a major concern in climate change discussions due to the potential for triggering cascading effects that could amplify global warming and its associated impacts.

3. The data in this chapter are from Our World in Data which provides emissions data harmonized across countries; see https://ourworldindata.org/.

4. It includes energy-related emissions from electricity generation for lighting, appliances, cooking, heating and cooling for both residential and commercial buildings.

5. CO_2 equivalent, abbreviated as CO_2e, is a standard unit for measuring carbon footprint. It equates the impact of different greenhouse gases on global warming relative to the same amount of CO_2. CO_2e is calculated by multiplying the mass of the greenhouse gas by its Global Warming Potential (GWP), a value that reflects its relative impact on warming the atmosphere compared to CO_2. This approach allows for a consistent measurement across various greenhouse gases.

6. Nationally Determined Contributions (NDCs) are central to the Paris Agreement. They represent the commitments made by each country to reduce national emissions and adapt to climate change. Under the Paris Agreement, every party must prepare, communicate and maintain successive NDCs that they aim to achieve, reflecting their highest possible ambition.

7. A factor that contributed to the significant disparity between the declines in production and emissions in 2020 was that most of the downturn occurred in the service sectors, which generally produce fewer emissions, while the operations of high emitting sectors, such as energy and heavy industry, largely persisted.

8. Note that, in our modern era, particularly in developed nations, regulations are designed to alleviate and control the negative effects of industrial waste.

9. Human capital refers to the economic value of a worker's experience and skills. It includes education, training, intelligence, skills, health, and other aspects valued by employers, such as loyalty and punctuality. Human capital is developed through education, training and healthcare and is an essential factor determining the productive capacity of an individual or work-force, which influence an entity's or a nation's economic growth.

10. A shadow price is the estimated price of a good or service for which no market price exists. It represents the implicit value of a resource that is not currently priced in the market. For instance, a shadow price would be the estimated monetary value of the ecological services provided by a natural resource, like the Amazon rainforest's capacity to absorb carbon dioxide. This price aims to reflect the true cost or value of resources that

are otherwise unpriced or under-priced in the market, helping to inform decision-making that takes into account the social and environmental costs or benefits associated with economic activities.

11. Note that there are disparities in how different economic models perceive and account for adverse natural climate impacts. For instance, computable general equilibrium models (GEM) such as the GEM-E3 used by the EC and Cambridge Econometrics E3ME, do not include functions for climate damage, making them unable to quantify the effects of the negative growth associated with escalating temperatures and natural disasters. In contrast, integrated assessment models capture the harmful effects of rising temperatures and associated climatic disruptions and show that, in the context of long-term economic growth, scenarios championing emissions reductions outperform "business-as-usual" ones (Catalano *et al.* 2021).

12. NIMBYism, an acronym for "Not In My Back Yard", refers to the phenomenon where residents oppose certain developments because they do not want them located near their homes, even though these developments are needed for the greater good of the community. Common targets of NIMBY opposition include infrastructure projects, shelters, and various industrial facilities.

13. Note that my reference to economists refers not to the whole profession, but rather a mindset that predominates within the majority.

14. The social discount rate tends to be a metric used in cost–benefit analyses, to account for the time value of money in public projects. It reflects the preference for receiving benefits now rather than in the future and is used to calculate the present value of future social benefits and costs. The rate is typically set lower than that used for private investment, to reflect the broader societal perspective that values future public benefits higher. The choice of a social discount rate affects investment decisions in public projects, with lower rates typically justifying investment in projects with long-term benefits, such as environmental conservation or public health initiatives.

15. Green budgeting refers to the process of incorporating environmental considerations into governments' and organizations' budgeting and financial decision-making processes. This would align financial flows, including expenditure and investments, with environmental sustainability objectives. The goal is to provide funding to support environmental policies, ensure efficient use of resources and promote sustainable development; see Bova (2021).

16. The G30 Working Group is part of the Group of Thirty, which is an international body consisting of leading figures from the financial and academic sectors. It aims to deepen the understanding of global economic and financial issues and explore the implications of decisions made in the public and private spheres. The G30 Working Group on Climate Change and Finance is a specialized arm of the Group of Thirty focused on issues related to climate change and its implications for global finance.

17. *Green* expenditures might also have a higher multiplier than *brown* ones; thus, they might contribute more to sustaining economic activity; see Batini *et al.* (2022).

18. See EEA greenhouse gases: data viewer; https://www.eea.europa.eu/data-and-maps/data/data-viewers/greenhouse-gases-viewer.

19. The European Commission is the executive branch of the European Union, responsible for proposing legislation, implementing decisions, upholding the EU treaties, and managing the day-to-day business of the EU. It operates similarly to a government cabinet, with Commissioners appointed to lead different areas such as trade, environment, or economics.

20. E-fuel or synthetic or electro-fuel, is a type of fuel that is manufactured using CO_2 and water, with electricity as the primary energy source. Ideally, the electricity should come from renewables to ensure the low carbon footprint of e-fuel. The process involves producing hydrogen through electrolysis and combining it with CO_2 in a synthesis process to produce hydrocarbons such as gasoline, diesel and jet fuel. Conventional internal combustion engine vehicles can run on e-fuel with no need for significant modifications.

21. Failure of a member state to comply with an EU directive triggers an EC infringement procedure that starts with a formal notice followed by a reasoned opinion or formal request for compliance. If non-compliance persists, the case may be referred to the European Court of Justice, which can impose a financial penalty. Non-compliance can also have political and diplomatic repercussions within the EU, which will affect the non-compliant country's relationships and influence. It can also have an effect on the member state's citizens and businesses who could miss out on the benefits and protections offered by the particular directive.

22. See Climate Action Tracker; https://climateactiontracker.org/countries/china/.

23. In addition, because it is difficult to provide tax incentives only to those who might otherwise not invest, tax incentives tend to be transitory, targeted at specific sectors and costly compared to the results they achieve, and often result in transfers to entities that would have invested anyway.

24. This is related, also, to the abandonment of nuclear power implemented by some European countries which has reduced the diversity of energy sources.

25. For a deeper dive into the DNSH concept and its application within the EU's sustainable finance framework, the technical guidance by the European Commission (ECCWG, 16 February 2021) and the explanations provided by the European Securities and Markets Authority (22 November 2023, ESMA30-379-2281) are valuable resources.

26. The CBAM regulation was approved by the European Parliament and the Council of Europe on 10 May 2023, was published in the EU's Official Journal on 16 May 2023.

27. See https://taxation-customs.ec.europa.eu/carbon-border-adjustment-mechanism_en.

28. The WTO MFN rule ensures non-discrimination among WTO member countries. Essentially, it requires that any favourable treatment granted to one member country (e.g. lower customs duties, reduced restrictions on imports, etc.) must immediately and unconditionally be extended to all other member countries.

29. For their contributions to atmospheric chemistry and understanding of ozone dynamics, in 1995, Molina and Rowland and Paul J. Crutzen were awarded the Nobel Prize in Chemistry, which highlights their significant contribution and knowledge about the formation and decomposition of ozone and the impact of their findings on environmental policy and global ecological stewardship.

30. HCFCs were compounds used as temporary replacements for CFCs and are less damaging to the ozone layer due to their hydrogen content, which allows more rapid breakdown and dispersal of gases in the atmosphere. However, HCFCs are GHGs and still contribute to ozone depletion. Under the Montreal Protocol, their use is being phased out and they are being replaced by more environmentally-friendly alternatives.

31. Article 5 of the Montreal Protocol refers to a provision that differentiates between developed and developing countries based on their levels of consumption of ozone-depleting substances. Countries that were consuming less than a certain amount of CFCs and other controlled substances at the time the Protocol was ratified are defined as Article 5 countries. These are typically developing nations and are subject to different requirements and timelines for the phase-out of these substances compared to developed, non-Article 5 countries. The Multilateral Fund was established to assist Article 5 countries in complying with the Protocol's provisions, ensuring they have access to the necessary resources and technologies for transition to safer alternatives.

32. Establishing a legal framework dedicated to climate and environmental goals not only formalizes these objectives but also enables the judicial pursuit of parties responsible for environmental harm and violations of environmental regulations. In recent years, litigation concerning environmental and climate-related matters has seen a significant increase.

References

Andor, M., A. Lange & S. Sommer 2022. "Fairness and the support of redistributive environmental policies". *Journal of Environmental Economics and Management* 114: 102682.

Barrage, L. 2020. "The fiscal costs of climate change". *AEA Papers and Proceedings* 110: 107–12.

Batini, N. *et al.* 2022. "Building back better: how big are green spending multipliers?". *Ecological Economics* 193(C).

Beiser-McGrath, L. & T. Bernauer 2019. "Could revenue recycling make effective carbon taxation politically feasible?". *Science Advances* 5(9).

Bergquist, M. *et al.* 2022. "Meta-analyses of fifteen determinants of public opinion about climate change taxes and laws". *Nature Climate Change* 12: 235–40.

Berners-Lee, M. 2010. *How Bad Are Bananas? The Carbon Footprint of Everything*. London: Profile Books.

Bova, E. 2021. "Green budgeting practices in the EU: a first review". European Commission Discussion Paper 140. https://economy-finance.ec.europa.eu/publications/green-budgeting-practices-eu-first-review_en.

Burningham, K. 2000. "Using the language of NIMBY: a topic for research, not an activity for researchers". *Local Environment* 5(1): 55–67.

Carney, M. 2015. "Breaking the tragedy of the horizon: climate change and financial stability". Bank of England. https://www.

bankofengland.co.uk/speech/2015/breaking-the-tragedy-of-the-horizon-climate-change-and-financial-stability.

Catalano, M. & L. Forni 2021. "Fiscal policies for a sustainable recovery and a green transformation". Policy research working paper, No. WPS 9799. Washington, DC: World Bank Group. http://documents.worldbank.org/curated/en/499301633704126369/Fiscal-Policies-for-a-Sustainable-Recovery-and-a-Green-Transformation.

Catalano, M., L. Forni & E. Pezzolla 2019. "Climate change adaptation: the role of fiscal policy". *Resource and Energy Economics* 59: 101111.

Climate Action Tracker 2023. *Warming Projections Global Update.* December 2023. https://climateactiontracker.org/documents/1187/CAT_2023-12-05_GlobalUpdate_COP28.pdf.

Colantone, I. *et al.* 2023. "The political consequences of green policies: evidence from Italy". *American Political Science Review* 18(1): 108–26.

Dabla-Norris, E. *et al.* 2023. "Public perceptions of climate mitigation policies: evidence from cross-country surveys". *IMF Staff Discussion Notes* (SDNs) SDN/2023/002.

Dasgupta, P. 2021. *The Economics of Biodiversity: The Dasgupta Review.* London: HM Treasury.

Devine-Wright, P. 2009. "Rethinking NIMBYism: the role of place attachment and place identity in explaining place-protective action". *Journal of Community & Applied Social Psychology* 19(6): 426–41.

Drouet, L., V. Bosetti & M. Tavoni 2022. "Net economic benefits of well-below 2°C scenarios and associated uncertainties". *Oxford Open Climate Change* 2(1): 1–7.

European Commission 2022. "REPowerEU: A plan to rapidly reduce dependence on Russian fossil fuels and fast forward the green transition". Press release, 18 May. https://ec.europa.eu/commission/presscorner/detail/en/ip_22_3131.

Douenne, T. & A. Fabre 2022. "Yellow vests, pessimistic beliefs, and carbon tax aversion". *American Economic Journal: Economic Policy* 14(1): 81–110.

Fransen, T. *et al.* 2023. "9 things to know about National Climate Plans (NDCs)". World Resources Institute, 7 December. https://www.wri.org/insights/assessing-progress-ndcs?utm_campaign=wridigest&utm_source=wridigest-2022-10-27&utm_medium=email.

Forni, L. *et al.* 2024. *The "Green Buildings" Directive: A Quantification of its Costs and Benefits in Two Italian Regions.* Mimeo.

G30 Working Group on Climate Change and Finance 2020. *Mainstreaming the Transition to a Net-Zero Economy.* https://group30.org/publications/detail/4791.

Gates, B. 2021. *How to Avert a Climate Crisis.* London: Penguin.

Gonguet, F. *et al.* 2021. "Climate-sensitive management of public finances – green PFM". IMF. https://www.imf.org/en/Publications/staff-climate-notes/Issues/2021/08/10/Climate-Sensitive-Management-of-Public-Finances-Green-PFM-460635.

Howarth, C. & B. Roberts 2018. "The role of the UK green deal in shaping pro-environmental behaviours: insights from two case studies". *Sustainability* 10(6).

IEA 2021. Greenhouse Gas Emissions from Energy Data Explorer. https://www.iea.org/data-and-statistics/data-tools/greenhouse-gas-emissions-from-energy-data-explorer.

IPCC 2021. *Climate Change 2021: The Physical Science Basis.* IPCC Assessment Report 6. https://www.ipcc.ch/report/ar6/wg1/.

IPCC 2022. *Climate Change 2022: Impacts, Adaptation and Vulnerability.* IPCC Assessment Report 6. https://www.ipcc.ch/report/ar6/wg2/.

Jagers, S. *et al.* 2021. "Bridging the ideological gap? How fairness perceptions mediate the effect of revenue recycling on public support for carbon taxes in the United States, Canada and Germany". *Review of Policy Research* 38(5): 529–54.

Mildenberger, M., E. Lachapelle & K. Harrison 2022. "Limited evidence that carbon tax rebates have increased public support for carbon pricing". *Nature Climate Change* 12: 121–2.

Nordhaus, W. 2013. *The Climate Casino.* New Haven, CT: Yale University Press.

Nowlin, M., K. Gupta & J. Ripberger 2020. "Revenue use and public support for a carbon tax". *Environmental Research Letters* 15(8): 084032.

OECD 2021. *Green Budgeting in OECD Countries*. Paris: OECD Publishing. https://doi.org/10.1787/acf5d047-en.

O'Neil, S. 2021. "Community obstacles to large scale solar: NIMBY and renewables". *Journal of Environmental Studies and Sciences* 11(1): 85–92.

Parry, I. 2021. "Five things to know about carbon pricing". International Monetary Fund. https://www.imf.org/en/Publications/fandd/issues/2021/09/five-things-to-know-about-carbon-pricing-parry.

Richman, B. 2002. "Mandating negotiations to solve the NIMBY problem: a creative regulatory response". *UCLA Journal of Environmental Law and Policy* 20: 223–36.

Sapir, A. 2021. "The European Union's carbon border mechanism and the WTO". Bruegel blog post, 19 July. https://www.bruegel.org/blog-post/european-unions-carbon-border-mechanism-and-wto.

Stern, N. *et al.* 2022. *Collaborating and Delivering on Climate Action through a Climate Club: An Independent Report to the G7*. London School of Economics and Political Science.

Stock, J. & D. Stuart 2021. "Emissions and electricity price effects of a small carbon tax combined with renewable tax credit extensions". Working Paper. https://scholar.harvard.edu/stock/publications/emissions-and-electricity-price-effects-small-carbon-tax-combined-renewable-tax.

The White House 2023. "One year in, President Biden's Inflation Reduction Act is driving historic climate action and investing in America to create good paying jobs and reduce costs". Fact Sheet, 16 August. https://www.whitehouse.gov/briefing-room/statements-releases/2023/08/16/fact-sheet-one-year-in-president-bidens-inflation-reduction-act-is-driving-historic-climate-action-and-investing-in-america-to-create-good-paying-jobs-and-reduce-costs/.

World Meteorological Organization (WMO) 2022. *Executive Summary. Scientific Assessment of Ozone Depletion*. GAW Report No. 278. Geneva: WMO.

Index